POTENTIAL ENERGY SURFACES

**molecular structure and
reaction dynamics**

POTENTIAL ENERGY SURFACES

molecular structure and reaction dynamics

David M. Hirst
Department of Chemistry,
University of Warwick, England

Taylor & Francis
London and Philadelphia
1985

UK   Taylor & Francis Ltd, 4 John St, London WC1N 2ET

USA   Taylor & Francis Inc., 242 Cherry St, Philadelphia, PA 19106–1906

British Library Cataloguing in Publication Data

Hirst, David M.
    Potential energy surfaces: molecular structure and reaction dynamics.
    1. Excited state chemistry   2. Molecules
    I. Title
    541.2'8   QD461.5

    ISBN 0-85066-275-3

Library of Congress Cataloging in Publication Data

Hirst, David M. (David Michael)
    Potential energy surfaces.

    Bibliography: p.
    Includes index.
    1. Excited state chemistry   2. Molecular structure.
I. Title
QD461.5.H57 1985   541.2'2   84-26751
ISBN 0-85066-275-3

Printed in Great Britain by Taylor & Francis (Printers) Ltd, Basingstoke, Hants.

CONTENTS

To Judy

PREFACE

The concept of a potential energy surface, which is based on the assumption that in a molecule, electronic motion and nuclear motion are separable, is one of the most important ideas in chemical physics. Much research effort is directed both to determining potential energy surfaces from spectroscopic data, by the methods of quantum chemistry or from molecular beam experiments, and to using such surfaces for dynamical calculations and the simulation of spectra. Although there are many specialist reviews on various aspects of potential energy surfaces and collections of such articles (e.g., K. P. Lawley (1980), *Potential Energy Surfaces* (Wiley) or *Faraday Discussions of the Chemical Society*, Volume 62), the author is not aware of an introductory text on this subject. This book is an attempt to bridge the gap between standard undergraduate texts on spectroscopy, quantum theory and bonding, and reaction kinetics, and more advanced review articles.

The aim of the book is to give an account of methods by which potential energy surfaces for small molecules can be determined and of the relationships between the potential surface and molecular spectra and dynamics at a level suitable for senior undergraduates and beginning graduate students. Spectroscopy plays a very important part in this book and it is assumed that the reader has a basic knowledge of the spectra and energy levels of simple molecules. We are not concerned here with details of spectra and their interpretation but in deriving information about the potential energy surface from molecular energy levels. A knowledge of molecular orbital theory and of the electronic structures of simple molecules is also assumed. Undergraduate courses usually place less emphasis on reaction dynamics so this topic is treated in more depth here.

I am very grateful to Professor I. M. Mills, who initially encouraged me to write the book, for critically reading the manuscript and suggesting many improvements. I also thank Professor T. J. Kemp and Dr D. J. Milton for reading parts of the manuscript and for making several useful comments. I thank Professor J. N. Murrell for permission to reproduce some contour plots from the SERC Potential Energy Surface Data Base, and Academic Press, The American Institute of Physics, Clarendon Press, Pergamon Press,

Plenum Press, The Royal Society of Chemistry, Springer–Verlag, Taylor & Francis and Van Nostrand Company for permission to reproduce figures from their publications. The authors have also kindly given their consent for this material to be reproduced here. The manuscript was prepared on the University of Warwick computer and photoset from the computer tape on a Linotron 202 at the University of Nottingham. Dr G. F. Paechter, of the Department of Mathematics, provided invaluable editorial and technical help without which the interfacing of these systems could not have been successfully achieved.

Finally, I am grateful to my wife for her encouragement and support during the writing of this book.

D. M. Hirst
Coventry, November 1984

CHAPTER 1

the separability of electronic and nuclear motion—the Born–Oppenheimer separation

1.1. Introduction

The concept of a potential energy surface plays a central role in our understanding of molecular phenomena. The formal definition of a potential energy surface involves a fairly complicated mathematical argument and will be given later but a qualitative discussion will be given here by way of introduction. The concept of a potential energy surface comes from the recognition that in a molecule the motion of the electrons is much faster than that of the nuclei and that these two types of motion can be separated. This separability of nuclear and electronic motion is known as the Born–Oppenheimer separation. We can think of the electrons as being subject to the electrostatic field of the stationary nuclei. The electronic wavefunction is obtained by solving the Schrödinger equation for a given nuclear configuration. On the other hand the nuclei are subject to a potential which is the sum of the nuclear repulsion and the average field due to the electrons. The wavefunction for the nuclei is thus obtained by solving a Schrödinger equation in which the nuclei are subject to a potential which is the sum of the electronic energy and the electrostatic potential of the nuclei. This potential, which is, of course, a function of the nuclear geometry, is the potential energy function or potential energy surface for nuclear motion. The potential energy function for a diatomic molecule is a function only of the internuclear separation and can be represented by the familiar potential energy curve. For a molecule containing N atoms, the potential energy function or potential energy surface is a function of $3N-6$ variables. Such functions cannot be represented graphically except for diatomic molecules. It is possible to portray the potential energy function for a polyatomic molecule only by constraining some of the variables. In the case of a triatomic molecule, for example, we may wish to consider collinear geometries only or perhaps keep one internuclear distance fixed and consider the variation in energy as the third atom is moved. The potential energy is then a function of two variables and can be represented as a three-dimensional model or in the form of a contour diagram. This gives rise to the concept of the potential energy

1

function being represented by a surface. Examples of several representations will be given in chapter 3. Because the potential energy function for a polyatomic molecule cannot be fully represented graphically, it is often referred to as a potential hypersurface.

There is clearly an intimate relationship between the potential energy surface for a molecule and its vibrational spectrum. The vibrational energy levels are obtained by solving the Schrödinger equation for the nuclear motion using the potential energy surface. Conversely one can obtain details of the potential energy surface from a detailed analysis of vibrational spectra. There is currently considerable interplay between theory and experiment, because for diatomic and small polyatomic molecules it is now possible to use the methods of quantum chemistry (see chapter 5) to make very accurate calculations of the electronic energy as a function of geometry. Direct comparison with experimental data can then be made by comparing the observed vibrational spacings with those calculated using the theoretical potential energy surfaces. Increasingly sophisticated spectroscopic techniques are yielding more and more information about molecular energy levels. These data can often only be interpreted satisfactorily in terms of accurate theoretical potential energy surfaces.

Chemistry is concerned with chemical reactions as well as with the properties of isolated molecules. Here again the potential energy surface plays a central role in our understanding of the dynamics of chemical reactions. At the most qualitative level, the activation barrier can be understood in terms of the variation of potential energy along the reaction coordinate. At a more detailed and quantitative level, the dynamics of a chemical reaction can be understood fully only in terms of motion of the nuclei on a potential energy surface (assuming that the Born–Oppenheimer separation is valid). Any theoretical treatment of the dynamics, whether classical, semiclassical or quantum, requires some form of potential energy surface. From molecular beam experiments, which are often coupled to spectroscopic techniques, we are obtaining increasingly detailed information about reaction dynamics, energy disposal and the quantum states of the products. From the data, deductions can often be made regarding the nature of the potential energy surface and comparison can be made with calculated potential surfaces and with the results of classical trajectory or scattering calculations. Also model calculations on empirical potential surfaces give valuable insight into the dynamics of many processes.

1.2. The Born–Oppenheimer Separation

In this section we give a formal definition of a potential energy surface in terms of the Born–Oppenheimer separation of nuclear and electronic motion. The validity of the separation will be examined in section 1.3 and in section 1.4 we give an outline of the methods to be used when the Born–Oppenheimer separation is not valid. The Schrödinger equation for the motion of the electrons and the nuclei in a molecule is of the form

$$[T_{nuc}(\boldsymbol{R}) + T_{el}(\boldsymbol{r}) + V_{nn}(\boldsymbol{R}) + V_{ne}(\boldsymbol{R},\boldsymbol{r}) + V_{ee}(\boldsymbol{r})]\Psi(\boldsymbol{R},\boldsymbol{r}) = E\Psi(\boldsymbol{R},\boldsymbol{r}). \quad (1.1)$$

$T_{nuc}(\boldsymbol{R})$ is the kinetic energy operator for the nuclei (at positions given by $\boldsymbol{R}$, relative to the centre of mass of the molecule), $T_{el}(\boldsymbol{r})$ is is the kinetic energy operator for the electrons (at position $\boldsymbol{r}$, relative to the centre of mass), $V_{nn}(\boldsymbol{R})$ is the electrostatic potential energy of the nuclei, $V_{ne}(\boldsymbol{R},\boldsymbol{r})$ is the potential energy of the electrons in the field of the nuclei and $V_{ee}(\boldsymbol{r})$ is the inter-electronic repulsion. Nuclei are much heavier than electrons and thus nuclear motion is very slow compared with electronic motion and it would seem to be a reasonable approximation to separate the two types of motion. This is the Born–Oppenheimer separation in which the total wavefunction $\Psi(\boldsymbol{R},\boldsymbol{r})$ is written as the product of an electronic wavefunction $\Phi(\boldsymbol{R},\boldsymbol{r})$ for a given nuclear configuration $\boldsymbol{R}$ and of a nuclear wavefunction $\chi(\boldsymbol{R})$:

$$\Psi(\boldsymbol{R},\boldsymbol{r}) = \Phi(\boldsymbol{R},\boldsymbol{r})\chi(\boldsymbol{R}). \quad (1.2)$$

The electronic wavefunction $\Phi(\boldsymbol{R},\boldsymbol{r})$ is the solution to the electronic Schrödinger equation

$$[T_{el}(\boldsymbol{r}) + V_{ne}(\boldsymbol{R},\boldsymbol{r}) + V_{ee}(\boldsymbol{r})]\Phi(\boldsymbol{R},\boldsymbol{r}) = W(\boldsymbol{R})\Phi(\boldsymbol{R},\boldsymbol{r}). \quad (1.3)$$

This equation describes the motion of the electrons in the field of the nuclei at positions $\boldsymbol{R}$. The wavefunction $\Phi(\boldsymbol{R},\boldsymbol{r})$ and the energy $W(\boldsymbol{R})$ depend on the nuclear coordinates $\boldsymbol{R}$ and the use of this equation is often called the clamped nuclei approximation. $\chi(\boldsymbol{R})$ is the solution to the nuclear Schrödinger equation

$$[T_{nuc}(\boldsymbol{R}) + V_{nn}(\boldsymbol{R}) + W(\boldsymbol{R})]\chi(\boldsymbol{R}) = E\chi(\boldsymbol{R}). \quad (1.4)$$

The terms $V_{nn}(\boldsymbol{R})$ and $W(\boldsymbol{R})$ represent the potential energy to which the nuclei are subject and it is the sum of these terms $U(\boldsymbol{R}) = V_{nn}(\boldsymbol{R}) + W(\boldsymbol{R})$ which is known as the potential energy function or potential energy surface. For a diatomic molecule it is simply a function of the internuclear distance and can be represented by a potential energy curve. Figure 1.1 indicates the typical form of a potential energy curve for a bound diatomic molecule. As the internuclear distance is increased the energy increases and asymptotically approaches the sum of the energies of the constituent atoms (or ions) in their respective states. The minimum in the curve corresponds to the equilibrium configuration for an internuclear separation of $R = R_e$. At shorter internuclear distances, the energy increases rapidly with decreasing R and approaches infinity as R tends to zero. The depth of the potential well represents the dissociation energy $\tilde{D}_e$ relative to the equilibrium

POTENTIAL ENERGY

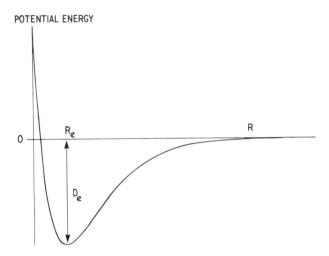

Figure 1.1. A typical potential energy curve for a diatomic molecule. R_e is the equilibrium bond length and D_e is the dissociation energy.

configuration. For a polyatomic molecule containing N atoms, the potential energy surface is a function of $3N-6$ variables. A function of more than two variables cannot be fully represented in graphical form and methods of presentation will be discussed in Chapter 3. Such potential energy surfaces are calculated on the assumption that the nuclei are moving slowly compared with the electrons and are known as adiabatic potential energy surfaces.

1.3. A More Detailed Look at the Born–Oppenheimer Separation

To examine the extent to which equation (1.2) is an approximation, we substitute into the full Schrödinger equation (1.1) to give

$$[T_{\text{nuc}}(\boldsymbol{R}) + T_{\text{el}}(\boldsymbol{r}) + V_{\text{nn}}(\boldsymbol{R}) + V_{\text{ne}}(\boldsymbol{R}, \boldsymbol{r}) + V_{\text{ee}}(\boldsymbol{r})]\Phi(\boldsymbol{R}, \boldsymbol{r})\chi(\boldsymbol{R})$$

$$= E\Phi(\boldsymbol{R}, \boldsymbol{r})\chi(\boldsymbol{R}). \quad (1.5)$$

The kinetic energy operators are of the form

$$T_{\text{nuc}}(\boldsymbol{R}) = -\frac{h^2}{8\pi^2} \sum_k \frac{1}{M_k} \nabla_k^2 \quad (1.6)$$

$$T_{\text{el}}(\boldsymbol{r}) = -\frac{h^2}{8\pi^2 m} \sum_i \nabla_i^2 \quad (1.7)$$

where the summation k runs over the nuclei (of mass M_k) and the summation i runs over the electrons (of mass m). Thus

$$T_{\text{nuc}}(\boldsymbol{R})\Phi(\boldsymbol{R},r)\chi(\boldsymbol{R}) = \Phi(\boldsymbol{R},r)\left[-\frac{h^2}{8\pi^2}\sum_k \frac{1}{M_k}\nabla_k^2\chi(\boldsymbol{R})\right]$$

$$+\chi(\boldsymbol{R})\left[-\frac{h^2}{8\pi^2}\sum_k \frac{1}{M_k}\nabla_k^2\Phi(\boldsymbol{R},r)\right]$$

$$-\frac{h^2}{8\pi^2}\sum_k \frac{2}{M_k}\{\nabla_k\Phi(\boldsymbol{R},r)\}\{\nabla_k\chi(\boldsymbol{R})\}. \quad (1.8)$$

Use of equations (1.3) and (1.4) then shows that the separation of the wavefunction as in equation (1.2) is only valid if the term

$$-\frac{h^2}{8\pi^2}\sum_k \frac{1}{M_k}[\chi(\boldsymbol{R})\nabla_k^2\Phi(\boldsymbol{R},r)+2\{\nabla_k\Phi(\boldsymbol{R},r)\}\{\nabla_k\chi(\boldsymbol{R})\}] \quad (1.9)$$

is equal to zero. These terms can be shown to be negligible in most cases. For example, in the first term, $\nabla_k^2\Phi(\boldsymbol{R},r)$ consists of second derivatives of the electronic wavefunction $\Phi(\boldsymbol{R},r)$ with respect to the nuclear coordinates and will be of the same order of magnitude as the derivative with respect to electronic coordinates $\nabla_i^2\Phi(\boldsymbol{R},r)$. However $h^2/(8\pi^2 m)\nabla^2\Phi(\boldsymbol{R},r)$ is of the same order of magnitude as the energy of one of the electrons. The first term in equation (1.9) is thus of the order of m/M_k of the electronic energy and can be neglected because nuclear masses are several thousand times larger than the electronic mass. Similarly, the second term can be neglected.

1.4. The Adiabatic Approximation

We discuss here a slightly different approach which will used in the treatment of non-adiabatic effects in spectroscopy and which will form the basis of our discussion of non-adiabatic transitions in reaction dynamics in chapter 6. Instead of simply assuming that the total wavefunction $\psi(\boldsymbol{R},r)$ can be written as the product of an electronic wavefunction $\Phi(\boldsymbol{R},r)$ and a nuclear wavefunction $\chi(\boldsymbol{R})$ as in equation 1.2, the total wavefunction $\psi(\boldsymbol{R},r)$ is expanded as a linear combination of such products over all electronic states j of the system

$$\psi(\boldsymbol{R},r) = \sum_j \chi_j(\boldsymbol{R})\Phi_j(\boldsymbol{R},r). \quad (1.10)$$

The functions $\Phi_j(\boldsymbol{R},r)$ are solutions of the electronic Schrödinger equation

1.3. This expression is substituted into the full Schrödinger equation 1.1. Multiplication by a function $\Phi_i(R,r)$ and integration with respect to the electronic coordinates r leads to the result

$$[T_{\text{nuc}} + U_i(R) - E]\chi_i(R) + \sum_j c_{ij}(R,P)\chi_j(R) = 0. \tag{1.11}$$

where $U_i(R)$ is the potential energy function for state i and

$$c_{ij}(R,P) = \sum_k \frac{1}{M_k}(A_{ij}^{(k)} \cdot P_k + B_{ij}^{(k)}). \tag{1.12}$$

The operator P_k is the nuclear momentum operator $-(ih/2\pi)\nabla_k$ and the terms $A_{ij}^{(k)}$ and $B_{ij}^{(k)}$ are defined by

$$A_{ij}^{(k)}(R) = \int \Phi_i^*(R,r)\left(-\frac{ih}{2\pi}\nabla_k\right)\Phi_j(R,r)\,d\tau \tag{1.13}$$

$$B_{ij}^{(k)}(R) = \int \Phi_i^*(R,r)\left(-\frac{h^2}{8\pi^2}\nabla_k^2\right)\Phi_j(R,r)\,d\tau. \tag{1.14}$$

For a stationary state the electronic wavefunction $\Phi_i(R,r)$ can be chosen to be a real function and the diagonal term $A_{ii}^{(k)}$ is equal to zero. Equation 1.11 can be rewritten as

$$[T_{\text{nuc}} + U_i(R) + B_{ii}(R) - E]\chi_i(R) = -\sum_{j\neq i} c_{ij}(R,P)\chi_j(R). \tag{1.15}$$

To solve equations 1.11 without approximation requires an infinite basis set for the expansion in equation 1.10. However the coupling terms c_{ij} are often only important between a small group of states Φ_j, or even just a pair of states. The expansion can then be truncated to a small group of basis functions. In the adiabatic approximation, the coupling terms c_{ij} between the different electronic states are neglected entirely and one has the equation

$$[T_{\text{nuc}} + U_i(R) + B_{ii}(R) - E]\chi(R) = 0 \tag{1.16}$$

for the nuclear wavefunction which differs from equation 1.4 by the presence of the term $B_{ii}(R)$ which is a small adiabatic correction to the potential energy surface. The total wavefunction is given as before by equation 1.2. If the terms c_{ij} are not negligible, there will be coupling between the electronic state i and other states and the Born–Oppenheimer separation is not valid.

Suggestions for Further Reading

M. Born and K. Huang (1954) *Dynamical Theory of Crystal Lattices*, Clarendon Press, Oxford.
H. Eyring and S. H. Lin (1974) in *Physical Chemistry, An Advanced Treatise*, ed. W. Jost, Vol. 6A, p. 121, Academic Press, New York.

G. Herzberg (1966) *Molecular Spectra and Molecular Structure,* Vol. III. *Electronic Spectra and Electronic Structure of Polyatomic Molecules,* Van Nostrand, Princeton.

J. C. Slater (1963) *Quantum Theory of Molecules and Solids,* Vol. 1. *Electronic Structure of Molecules,* McGraw-Hill, New York.

CHAPTER 2

diatomic molecules

2.1. Model Potential Energy Functions

There are two methods by which the potential energy curve for a molecule can be obtained. From the vibrational spectrum it is possible to calculate the turning points for the vibrational motion by the semi-classical Rydberg–Klein–Rees procedure. These turning points define the potential curve. This approach will be outlined in section 2.3. Alternatively it is possible to calculate theoretically the potential curve by high-quality *ab initio* quantum chemistry methods to be discussed in detail in chapter 5. It is only within the past decade that this approach has yielded sufficiently accurate results and this method is as yet generally limited to molecules containing light atoms. Both of these approaches require a considerable amount of effort and neither yields analytical forms for the potential curve. Thus many attempts have been made to develop functions which give a reasonable representation of the potential. The Schrödinger equation can then be solved directly or, if that is not possible, expressions for the energy levels obtained by perturbation theory. We discuss some of these model potentials in this section and consider the relationship between the potential function and the vibrational spectrum.

In order to obtain the nuclear energy levels E, we have to solve the nuclear Schrödinger equation

$$[T_{\text{nuc}}(\boldsymbol{R}) + V_{\text{nn}}(\boldsymbol{R}) + W(\boldsymbol{R})]\chi(\boldsymbol{R}) = E\chi(\boldsymbol{R}) \tag{2.1}$$

which, for a diatomic molecule, can be rewritten as

$$\sum_{k=1}^{2} \frac{1}{M_k} \nabla_k^2 \chi(\boldsymbol{R}) + \frac{8\pi^2}{h^2}[E - U(\boldsymbol{R})]\chi(\boldsymbol{R}) = 0. \tag{2.2}$$

Different energies E and wavefunctions $\chi(\boldsymbol{R})$ are obtained for each set of nuclear quantum numbers for a given electronic state. The function $U(\boldsymbol{R})$, the potential energy function, is given by

$$U(\boldsymbol{R}) = V_{\text{nn}}(\boldsymbol{R}) + W(\boldsymbol{R}) \tag{2.3}$$

9

which we recognize as depending only on the internuclear separation R. Transformation of equation (2.2) to spherical polar coordinates (R, θ, ϕ) gives

$$\frac{1}{R^2}\frac{\partial}{\partial R}\left(R^2\frac{\partial \chi}{\partial R}\right) + \frac{1}{R^2 \sin \theta}\frac{\partial}{\partial \theta}\left(\sin \theta \frac{\partial \chi}{\partial \theta}\right)$$

$$+ \frac{1}{R^2 \sin^2 \theta}\frac{\partial^2 \chi}{\partial \phi^2} + \frac{8\pi^2 \mu}{h^2}[E - U(R)]\chi = 0, \qquad (2.4)$$

where μ is the reduced mass $M_1 M_2/(M_1 + M_2)$. The wavefunction $\chi(R, \theta, \phi)$ can be written as the product of a radial function $\psi(R)$ and of an angular function $Y_{JM}(\theta, \phi)$ thus

$$\chi(R) \equiv \chi(R, \theta, \phi) = \psi(R) Y_{JM}(\theta, \phi), \qquad (2.5)$$

where J and M are rotational quantum numbers. The Schrödinger equation (2.4) can be separated by standard techniques (see Pauling and Wilson 1935) into a radial equation for $\psi(R)$ and equations for the angular functions constituting $Y_{JM}(\theta, \phi)$. The solutions to the angular equation are the familiar spherical harmonics $Y_{JM}(\theta, \phi)$ which are essentially products of an associated Legendre function $P_J^{|M|}(\cos \theta)$ and an exponential function $1/(2\pi)^{\frac{1}{2}}\exp(iM\phi)$. The resulting radial equation is

$$\frac{1}{R^2}\frac{d}{dR}\left(R^2\frac{d\psi}{dR}\right) + \left[\frac{8\pi^2 \mu}{h^2}[E - U(R)] - \frac{J(J+1)}{R^2}\right]\psi = 0. \qquad (2.6)$$

Solution of this equation yields the vibration–rotation energy levels $E_{v,J}$ and the corresponding vibrational wavefunctions $\psi_{v,J}$. The term $J(J+1)/R^2$ is an effective vibrational potential energy, derived from the rotational kinetic energy, associated with the centrifugal force due to the rotational angular momentum characterized by the quantum number J.

2.1.1. The Harmonic Potential

The simplest model for the vibration of a diatomic molecule is to assume a Hooke's law model in which the restoring force is proportional to the displacement from equilibrium. This force law corresponds to a quadratic potential function

$$U(R) = \tfrac{1}{2}k(R - R_e)^2 \qquad (2.7)$$

where k is known as the force constant. This model is clearly of limited validity as the parabolic potential is incapable of representing the asymptotic behaviour required for dissociation. Nevertheless it forms a reasonable first approximation in the vicinity of R_e as shown in figure 2.1. If rotation is ignored, i.e. $J = 0$, and we substitute $\psi(R) = S(R)/R$ and the harmonic

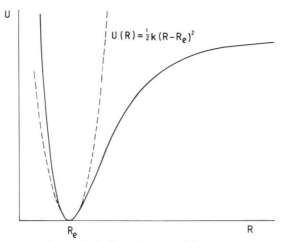

Figure 2.1. Comparison of a typical diatomic potential energy curve with the harmonic potential.

potential of equation 2.7 into equation 2.6, we obtain the Schrödinger equation for the one-dimensional harmonic oscillator. This equation is readily solved to give the familiar energy level expression

$$E_v = h\nu_e(v + \tfrac{1}{2}) = hc\omega_e(v + \tfrac{1}{2}), \tag{2.8}$$

where $\nu_e = 1/(2\pi)(k/\mu)^{\frac{1}{2}}$ and v is the vibrational quantum number which can take values $v = 0, 1, 2, \ldots$. In equation 2.8, if the vibration frequency ν_e is in hertz and the vibration wavenumber ω_e is in units of cm^{-1}, the energy E_v will be in joules. Solution of equation 2.6 is more complicated when the rotational contribution to the potential is included. Full details are given by Pauling and Wilson (1935) who show that the result is approximately

$$E_{v,J} = hc[\omega_e(v + \tfrac{1}{2}) + B_e J(J + 1) - D_e J^2(J + 1)^2] \tag{2.9}$$

where B_e is the rotational constant which can be written in terms of the moment of inertia I_e at the equilibrium bond length R_e ($I_e = \mu R_e^2$),

$$B_e = \frac{h}{8\pi^2 c I_e} = \frac{h}{8\pi^2 c \mu R_e^2}. \tag{2.10}$$

D_e is the centrifugal distortion constant which is related to B_e and ω_e by the expression

$$D_e = 4B_e^3/\omega_e^2. \tag{2.11}$$

Units of cm^{-1} are usually used for the constants ω_e, B_e and D_e in equation 2.9. This model predicts that the vibrational energy levels are equally spaced and consideration of the selection rules (assuming that the dipole moment varies linearly with internuclear distance) indicates that $\Delta v = \pm 1$. Thus the pure vibrational spectrum would be expected to consist of one single band. This is contrary to experiment, for vibrational spectra of diatomic molecules consist typically of one very intense band (the fundamental band) with a number of weaker bands at frequencies which are approximately (but rather less than) integral multiples of the fundamental frequency. These weaker bands can be interpreted as overtone bands arising from transitions with $\Delta v = \pm 2, \pm 3, \pm 4$, etc. Such transitions are, in fact, allowed when quadratic and higher terms are included in the expansion of the dipole moment as a function of internuclear distance. This phenomenon is known as electrical anharmonicity. However experiment indicates that the separation between successive vibrational levels decreases as the quantum number v increases. The harmonic oscillator model is unable to account for this. Indeed qualitative considerations indicate that the harmonic potential will be increasingly unsatisfactory as the potential energy increases.

2.1.2. *Anharmonic Potentials*

In order to account for the observed pattern of energy levels we have to use an anharmonic potential. This is known as mechanical anharmonicity. There are two possible approaches to obtaining anharmonic potential functions which yield calculated vibrational energy levels which are in better agreement with experiment than those given by the harmonic oscillator model. One approach is to modify the quadratic Hooke's law potential by the addition of cubic and quartic terms. In doing this we are essentially writing the potential as a Taylor series expansion in terms of the displacement $(R - R_e)$ from equilibrium. In order to establish notation that will be consistent with that used for polyatomic molecules in chapter 4, we write

$$U(R) = \frac{1}{2!} f_2 (R - R_e)^2 + \frac{1}{3!} f_3 (R - R_e)^3 + \frac{1}{4!} f_4 (R - R_e)^4, \qquad (2.12)$$

where f_2 is the harmonic force constant. The cubic and quartic force constants f_3 and f_4 are very much smaller than f_2. The Schrödinger equation cannot be solved exactly for this potential but the effect of the additional terms can be calculated by perturbation theory. Alternatively, one can look for an analytical function which has the correct sort of behaviour (as indicated in figure 1.1) and for which the Schrödinger equation can be solved.

If cubic and quartic terms are added to the quadratic potential function one obtains, in the absence of rotation, the following expression for the energy levels

$$E = hc\omega_e (v + \tfrac{1}{2}) - hcx(v + \tfrac{1}{2})^2 + hcy(v + \tfrac{1}{2})^3 \qquad (2.13)$$

where we have replaced the conventional symbols $\omega_e x_e$ and $\omega_e y_e$ by x and y

to adopt notation analogous to that used for polyatomic molecules. Alternatively the vibrational energy levels can be expressed as term values $G(v)$, which are usually expressed in units of cm^{-1},

$$G(v) = \omega_e(v+\tfrac{1}{2}) - x(v+\tfrac{1}{2})^2 + y(v+\tfrac{1}{2})^3. \tag{2.14}$$

In these expressions $x \ll \omega_e$ and $y \ll x$ and thus the separation between adjacent levels does indeed decrease as v increases.

In addition to the contribution to the transition intensity for values of Δv other than ± 1 arising from the presence of quadratic and higher terms in the dipole moment function, there is also a contribution arising from the use of an anharmonic potential. The wavefunction ψ_v for a level with quantum number v can be expanded in terms of harmonic oscillator wavefunctions ψ_k^0

$$\psi_v = \sum_{k=0}^{\infty} a_k \psi_k^0. \tag{2.15}$$

Hence there will be a non-zero contribution to the transition moment integral

$$\int \psi_{v'} \mu_x \psi_v \, dx$$

for cases other than $\Delta v = \pm 1$.

A very popular empirical potential function which has almost the correct behaviour is that suggested by Morse,

$$U(R) = \tilde{D}_e[1 - \exp\{-a(R-R_e)\}]^2 \tag{2.16}$$

where $\tilde{D}_e$ is the dissociation energy (in energy units) relative to the equilibrium configuration at R_e and the constant a is equal to $[k/(2\tilde{D}_e)]^{\frac{1}{2}}$. The potential can thus be fitted to values of $\tilde{D}_e$, R_e and k (the force constant). It does not become infinite at $R = 0$ but it does become very large and this region of the potential is not very important. The Schrödinger equation can be solved for this potential (see Pauling and Wilson (1935) for details) to yield the following energy expression

$$E_{v,J} = hc\omega_e(v+\tfrac{1}{2}) - hcx(v+\tfrac{1}{2})^2 + hcB_eJ(J+1)$$
$$- hcD_eJ^2(J+1)^2 - hc\alpha_e(v+\tfrac{1}{2})J(J+1). \tag{2.17}$$

This energy level expression contains the first two terms obtained from perturbation theory with the more general expansion of equation (2.12). We also get the rigid rotor rotational energy and the contribution from centrifugal distortion. In addition there is the term $\alpha_e(v+\tfrac{1}{2})J(J+1)$, a vibration–rotation interaction, which takes into account the change in the average moment of inertia due to vibration and the consequent change in rotational energy. Although the Morse function is not capable of giving a completely satisfactory fit to the actual potential curve over the whole range of internuclear distances, it does give a good fit in the region of R_e and has been very widely used. Cubic force constants derived from the Morse potential are usually in good agreement with those obtained from spectral analysis but agreement for the quartic force constants is usually rather poor. Repulsive potential curves can be represented by an anti-Morse function of the form

$$U(R) = \tfrac{1}{2}\tilde{D}_e[1+\exp\{-a(R-R_e)\}]^2. \tag{2.18}$$

An alternative procedure is to relate the empirical spectroscopic constants, obtained by fitting the observed frequencies to equation 2.17, to the coefficients f_2, f_3 and f_4 in the expression for the potential function as given in equation 2.12. The harmonic force constant f_2 is related to ω_e. The cubic force constant f_3 can be deduced from the values of ω_e, B_e and the vibration–rotation interaction constant α. Having obtained the cubic force constant, the quartic force constant f_4 can be calculated from ω_e, x and the cubic force constant. Full details are given by Mills (1974). The calculation of f_2, f_3 and f_4 does not constitute a determination of the complete potential energy function for the molecule because, although a function of the form of equation 2.12 is capable of giving a good fit to the potential function in the region of the first few vibrational levels, it will not behave correctly for large or small values of R.

Many empirical potential functions have been proposed. Murrell and Sorbie (1974) have shown that a modified version of the Rydberg potential

$$U(R) = -\tilde{D}_e[1+\beta(R-R_e)]\exp\{-\beta(R-R_e)\}, \tag{2.19}$$

where $\beta = (k/\tilde{D}_e)^{\frac{1}{2}}$, is capable of giving a very good fit to spectroscopic RKR potentials. Murrell and Sorbie increased the flexibility of the Rydberg potential by replacing the linear function by a cubic function thus

$$U(R) = -\tilde{D}_e[1+a_1x+a_2x^2+a_3x^3]\exp(-\gamma x), \tag{2.20}$$

where $x = R-R_e$ and a_1, a_2, a_3, and γ are adjustable parameters. Extended Rydberg potential functions have been derived by Huxley and Murrell (1983) for the ground states and some excited states of all the neutral diatomic molecules made up from the atoms H to Cl for which spectroscopic constants are given by Huber and Herzberg (1979). The extended Rydberg function is very successful in the bonding region of the potential curve. However it is not satisfactory for the long-range part of the potential which depends on R^{-6} or for parts of the potential which dissociate to ions.

2.2. Dunham's Method

Rather than seeking an empirical formula, Dunham has pursued a more general approach based on the expansion of the potential in powers of $(R-R_e)$ in the neighbourhood of the potential minimum. He chose an effective potential of the form

$$U = a_0 x^2(1 + a_1 x + a_2 x^2 + \dots) + B_e J(J+1)(1 - 2x + 3x^2 - 4x^3 + \dots), \quad (2.21)$$

where $x = (R - R_e)/R_e$. Using the semi-classical Wentzel–Kramers–Brillouin method, the energy levels can be written in the form

$$F_{vJ} = \sum_{l,j} Y_{lj}(v + \tfrac{1}{2})^l [J(J+1)]^j, \quad (2.22)$$

where l and j are summation indices and Y_{lj} are coefficients which depend on molecular constants and are related to the coefficients a_i in the potential function. Generally the ratio B_e/ω_e is small and the coefficients Y_{lj} can be related to the usual spectroscopic constants to yield the following expression for the term values

$$F_{vJ} = \omega_e(v + \tfrac{1}{2}) - x(v + \tfrac{1}{2})^2 + y(v + \tfrac{1}{2})^3 + z(v + \tfrac{1}{2})^4$$
$$+ B_v J(J+1) - D_v J^2(J+1)^2 + H_v J^3(J+1)^3, \quad (2.23)$$

where $B_v = B_e - \alpha_e(v + \tfrac{1}{2}) + \gamma_e(v + \tfrac{1}{2})^2 + \dots$. The parameters a_i in the potential function can be derived from differences between F_{vJ} (Gordy and Cook 1970).

2.3. The Rydberg–Klein–Rees (RKR) Procedure for the Determination of Diatomic Potential Functions

The potential functions discussed in the previous sections suffer from the disadvantage that an algebraic form is assumed for the potential curve. Although such a function may give an excellent fit for the lower vibrational levels, it may well be in serious error for the high vibrational levels. The Dunham potential assumes a polynomial expansion with respect to the equilibrium separation R_e and such an expansion cannot converge for very large values of R.

An alternative method for obtaining the potential energy curve is to use the experimentally determined vibration–rotation energy levels and to calculate from them the turning points of the vibrational motion R_{min} and R_{max} by a semi-classical method. The calculated turning points define the potential energy curve. This method was first proposed by Rydberg . Klein modified the method to circumvent difficulties in the graphical integration and Rees showed that use of an analytical expansion for the energy levels enabled one to obtain the turning points analytically. The following account is based on those of Vanderslice *et al.* (1959) and Mason and Monchick (1967).

In the quantum theory of Wilson and Sommerfeld quantization is introduced through action integrals. Hamilton's equations of motion are set up and solved in terms of generalized coordinates $q_1, \dots, q_{3n}$ and the canonically conjugate momenta $p_1, \ldots, p_{3n}$

$$p_i = -\frac{\partial H}{\partial q_i} \tag{2.24}$$

where H is the Hamiltonian. The action integral is defined as

$$\oint p_k \, dq_k$$

and in the Wilson–Sommerfeld theory the only classical orbits which are allowed as stationary states are those for which the action integral is equal to $n_k h$, where n_k is an integer.

In the RKR procedure for a diatomic molecule, the action integral I is written in terms of the radial momentum p_R

$$I = \oint p_R \, dR. \tag{2.25}$$

The radial momentum can be expressed in terms of the constant total energy U and the effective potential energy $V_{eff}(R)$ thus

$$U = \frac{p_R^2}{2\mu} + V_{eff}(R), \tag{2.26}$$

where μ is the reduced mass. The action integral is therefore given by

$$I = (2\mu)^{\frac{1}{2}} \oint [U - V_{eff}(R)]^{\frac{1}{2}} \, dR \tag{2.27}$$

which is set equal to $h(v + \frac{1}{2})$

$$I = h(v + \tfrac{1}{2}) = (2\mu)^{\frac{1}{2}} \oint [U - V_{eff}(R)]^{\frac{1}{2}} \, dR. \tag{2.28}$$

A second relationship between the potential energy function and an experimental observable is obtained from quantization of the rotational motion. The effective potential is written in terms of the actual potential $V(R)$ and the centrifugal potential in terms of the momentum p_θ

$$V_{eff}(R) = V(R) + \frac{p_\theta^2}{2\mu R^2} \tag{2.29}$$

as in equation 2.6. The rotational energy E_{rot} is equal to the quantity $p_\theta^2 / 2\mu R^2$ averaged over the vibrational period. Performing the average and replacing p_θ^2 by the quantum mechanical expression $(h^2/4\pi^2)J(J+1)$ gives

$$\begin{aligned}
E_{rot} &= \frac{h^2}{8\pi^2 \mu} J(J+1) \overline{\left(\frac{1}{R^2}\right)}_v \\
&= \frac{h^2}{8\pi^2 \mu} J(J+1) \frac{1}{\tau_v} \oint \frac{dt}{R^2} = \frac{h^2}{8\pi^2 \mu} J(J+1) \left(\frac{\mu}{\tau_v}\right) \oint \frac{dR}{R^2 p_R}, \tag{2.30}
\end{aligned}$$

where τ_v is the period of vibration. Substitution for p_R from equation 2.26

and replacement of E_{rot} by $hcB_vJ(J+1)$ gives

$$B_v = \frac{h}{8\pi^2 c \tau_v} \frac{1}{(2\mu)^{\frac{1}{2}}} \oint \frac{1}{[U-V_{eff}(R)]^{\frac{1}{2}}} \frac{dR}{R^2}. \qquad (2.31)$$

The potential can be derived by inserting the experimental values for the total energy U, the period of vibration τ_v and the rotational constant B_v into equations 2.28 and 2.31 and adjusting $V(R)$ to fit. Direct graphical integration of equation 2.31 is not very accurate because the integrand becomes infinite at the classical turning points. Klein showed that Rydberg's original method could be modified to obtain the turning points directly in terms of an auxiliary function S

$$S(U, K) = \frac{1}{(2\pi^2\mu)^{\frac{1}{2}}} \int_0^{I'} [U-E(I, K)]^{\frac{1}{2}} \, dI, \qquad (2.32)$$

where $E(I, K)$ is the vibration–rotation energy expressed in terms of the action integral I and K (which is related to the square of the angular momentum by $K = p_\theta^2/2\mu$). The upper limit of the integration I' is the value of I for which $E(I, K)$ equals U. It can be shown that S is one half of the area between the curve $V_{eff}(R)$ defining the effective potential and a line of constant energy U.

$$S(U, K) = \frac{1}{2} \int_{R_{min}}^{R_{max}} [U-V_{eff}(R)] \, dR = \frac{1}{2} \int_{R_{min}}^{R_{max}} \left(U-V(R)-\frac{K}{R^2}\right) dR. \qquad (2.33)$$

Thus

$$\frac{\partial S}{\partial U} = \frac{1}{2}(R_{max}-R_{min}) \qquad (2.34)$$

and

$$\frac{\partial S}{\partial K} = \frac{1}{2}\left(\frac{1}{R_{max}}-\frac{1}{R_{min}}\right) \qquad (2.35)$$

and the turning points can be obtained for a given observed energy U.

It was shown by Rees that the numerical integration of equation 2.33 followed by numerical differentiation of equations 2.34 and 2.35 to give $(R_{max}-R_{min})$ and $(1/R_{max}-1/R_{min})$ can be avoided if the energy $E(I, K)$ can be written as a quadratic in $(v+\frac{1}{2})$

$$E(I, K) = hc[\omega(v+\tfrac{1}{2})-x(v+\tfrac{1}{2})^2-\alpha(v+\tfrac{1}{2})J(J+1)$$
$$+ BJ(J+1)-DJ^2(J+1)^2+ \ldots]. \qquad (2.36)$$

Although one such expression is not usually capable of covering the whole experimental range, use of several quadratics enables the entire range to be covered.

The RKR method has been shown to be remarkably accurate despite being a first-order semi-classical approximation. Near the potential minimum the first-order treatment gives exact results. As the dissociation limit is reached, the motion becomes more classical and therefore a semi-classical approach should be more accurate in this region. The RKR method is now well established as a method for obtaining the potential energy function for a diatomic molecule, in numerical form, from the observed vibration–rotation energy levels.

2.4. Calculation of Vibration–Rotation Levels from the Potential Function for a Diatomic Molecule

The radial Schrödinger equation can be solved analytically for only a limited number of potential energy functions. In order to fit a numerical potential curve obtained from RKR or *ab initio* calculations it is necessary to use a flexible function, such as the extended Rydberg function proposed by Murrell and Sorbie (1974), which contains several parameters. Having obtained such a function one wishes to know how good it is in predicting the vibration–rotation spectrum. This can be done most satisfactorily by calculating the energy levels directly by a numerical method and comparing them with the experimental values. It is necessary to obtain solutions to the Schrödinger equation with an accuracy which is comparable to that with which the potential curve has been determined. This is best done by solving the equation by a numerical method proposed by Cooley (1961). Tests of the procedure have been described by Cashion (1963) and computer programs are available for this task (LeRoy 1979).

The centrifugal term in equation 2.6 is modified to include the contribution from Λ, the component of the electronic orbital angular momentum along the internuclear axis. The equation can be simplified by the substitution

$$\psi(R) = \frac{1}{R} S(R) \qquad (2.37)$$

to give

$$\frac{d^2 S}{dR^2} + \left[\frac{8\pi^2 \mu}{h^2} [E - U(R)] - \frac{[J(J+1) - \Lambda^2]}{R^2} \right] S = 0. \qquad (2.38)$$

It is not necessary to specify the potential function analytically. Numerical interpolation can be used to generate the potential at values of R intermediate between the calculated points.

2.5. The Term Manifold of Electronic States: Wigner–Witmer Rules

In the previous sections we have been concerned with the potential energy curve for a diatomic molecule in a particular electronic state and with the relationship between the potential energy curve and the vibration–rotation energy levels. This section is concerned, at a more qualitative level, with the set, or manifold, of molecular electronic states that can be formed from separated atoms in given electronic states. This will be discussed in terms of the rules derived by Wigner and Witmer. These rules do not give the form of the potential curves but often reasonable qualitative predictions can be made about the nature of the curve from a knowledge of the atomic states and the orbital occupancy at the dissociation limit. We shall assume that spin–orbit coupling is small and that Russell–Saunders coupling is a valid approximation. Thus orbital angular momentum and spin angular momentum will be separately coupled together. An alternative formulation is available for heavier atoms for which spin–orbit coupling is large.

We discuss the case of unlike atoms first and start with a discussion of the orbital angular momentum. The resultant orbital angular momentum for each atom is quantized and its magnitude is given by the quantum number L. Let L_A and L_B be the orbital angular momentum quantum numbers for atoms A and B. Molecules do not have spherical symmetry and the total orbital angular momentum is not well defined. However for a diatomic molecule, the component along the internuclear axis is well defined and is given by $\Lambda h/2\pi$ where Λ is a quantum number which can take values 0, ±1, ±2, ±3, etc. Molecular states are designated Σ, Π, Δ, Φ, etc. for values of $|\Lambda| = 0,1,2,3$, etc. When atoms A and B approach, L_A and L_B have $2L_A+1$ and $2L_B+1$ components respectively along the internuclear axis. These components are given by $M_A h/2\pi$ and $M_B h/2\pi$ where M_A and M_B are the appropriate atomic quantum numbers. M_A has integral values ranging from $-L_A$ to L_A. The allowed values of Λ are given by

$$\Lambda = M_A + M_B. \tag{2.39}$$

Thus any allowed orientation of L_A can be combined with any allowed orientation of L_B. The energy of the resultant electronic state depends on $|\Lambda|$ and provided that $M_A + M_B \neq 0$, we get a pair of degenerate states from the combinations $M_A + M_B$ and $-M_A - M_B$.

We consider the combination of an atom in an S state ($L_A = 0$) with one in a P state ($L_B = 1$). For the S state, since $L_A = 0$, $M_A = 0$ and we see that the angular momentum vector of the atom in the P state can have three allowed orientations with respect to the internuclear axis (see figure 2.2). In this case there are three allowed combinations of M_A and M_B, namely/p $\Lambda = -1, 0, 1$. The state with $|\Lambda| = 1$ is a doubly degenerate Π state and arises from the combinations $M_A = 0$, $M_B = 1$ and $M_A = 0$, $M_B = -1$. The

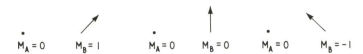

Figure 2.2. Allowed orientations of angular momentum vectors L_A and L_B for the combination of an atom in an S state with an atom in a P state.

state with $\Lambda = 0$ is a non-degenerate Σ state arising from the combination of $M_A = 0$ with $M_B = 0$.

Six states arise from the combination of two atoms each in a P state. Each atom has M values of $-1, 0, 1$ so $M_A + M_B$ can have values $-2, -1, 0, -1, 0, 1, 0, 1, 2$. Thus there is one Δ state with $|\Lambda| = 2$ from $M_A + M_B = 2, -2$, two Π states with $|\Lambda| = 1$ from $M_A + M_B = 1, -1$ and three Σ states with $\Lambda = 0$. Vector diagrams (figure 2.3) can be drawn to illustrate the allowed combinations of the angular momentum vectors.

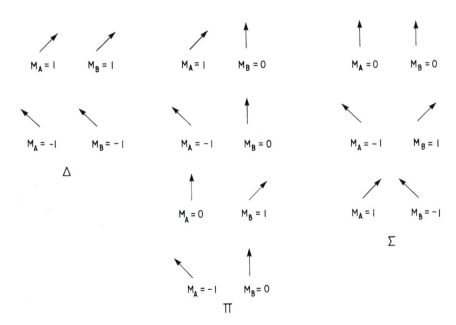

Figure 2.3. Allowed orientations of angular momentum vectors L_A and L_B for the combination of two atoms each in a P state.

Two of the Σ states arise from the combinations $M_A = 1$, $M_B = -1$ and $M_A = -1$, $M_B = 1$ and are degenerate in zero order. Examination of the wavefunctions for the allowed combinations shows that one combination is invariant with respect to reflection in a plane containing the nuclei and is therefore of symmetry Σ^+. The other combination changes sign under this operation and corresponds to a Σ^- state. A more complete treatment gives splitting of the Σ^+ and Σ^- states.

In our first example we obtained one Σ state whereas the second example yielded three Σ states. In fact one always obtains an odd number of Σ states. One occurs on its own and may be Σ^+ or Σ^- and the others occur in pairs of Σ^+ and Σ^-, as in the second example. In order to determine whether the single state is Σ^+ or Σ^-, one has to consider the parity of the constituent atoms. The parity is given by the sum of the individual orbital angular momentum quantum numbers $\sum_i l_{iA}$ for electrons outside closed subshells. Thus for the S state in Li arising from the configuration $1s^2 \, 2s^1$, $\sum l_{iA} = 0$ and the parity is even, whereas for the S state in N arising from the configuration $1s^2 \, 2s^2 \, 2p^1_{+1} \, 2p^1_{-1} \, 2p^1_0$, $\sum l_{iA} = 3$ and the parity is odd. If, for the combination of two atoms, the sum

$$L_A + L_B + \sum l_{iA} + \sum l_{iB}$$

is even then the state is Σ^+. If the sum is odd, then the state is Σ^-. Table 2.1 gives the molecular states arising from combinations of S, P and D atomic states. The subscripts g and u refer to states of even and odd parity for the atom. A more complete table is given by Herzberg (1950).

Each of the atoms is characterized by a spin quantum number S and by the vector model one obtains the following values for the resultant spin angular momentum

$$S = S_A + S_B, \ S_A + S_B - 1, \ S_A + S_B - 2, \ldots, \ |S_A - S_B|. \tag{2.40}$$

Table 2.2 contains the allowed combinations for singlets ($S = 0$), doublets ($S = 1$) and triplets ($S = 2$). For two states correlating with a given dissociation limit, the state with the highest resultant spin S lies lowest.

When two atoms with quantum numbers (L_A, S_A) and (L_B, S_B) combine, any of the allowed spin states from table 2.2 can combine with any of the allowed symmetries from table 2.1. Thus use of the rules outlined above enables one to determine how many states correlate with atoms in states defined by (L_A, S_A) and (L_B, S_B) and the multiplicities and symmetries of the allowed states.

If we consider the combination of two like atoms in the same electronic state, an additional complication arises from the symmetry of the wavefunction with respect to inversion of the electrons at the centre of symmetry. Table 2.3 gives details for a limited number of atomic states. Further details can be found in Herzberg (1950).

In the case of two like atoms in different electronic states, each of the states occurring for unlike atoms (as given in table 2.1) occurs twice, once with g symmetry and once with u symmetry. Thus if we consider the

Potential Energy Surfaces

Table 2.1. Molecular states obtained by combining two atomic states

States of separated atoms	Molecular states
$S_g + S_g$ or $S_u + S_u$	Σ^+
$S_g + S_u$	Σ^-
$S_g + P_g$ or $S_u + P_u$	Σ^-, Π
$S_g + P_u$ or $S_u + P_g$	Σ^+, Π
$S_g + D_g$ or $S_u + D_u$	Σ^+, Π, Δ
$S_g + D_u$ or $S_u + D_g$	Σ^-, Π, Δ
$P_g + P_g$ or $P_u + P_u$	Σ^+ (2), Σ^-, Π (2), Δ
$P_g + P_u$	Σ^+, Σ^- (2), Π (2), Δ
$P_g + D_g$ or $P_u + D_u$	Σ^+, Σ^- (2), Π (3), Δ (2), Φ
$P_g + D_u$ or $P_u + D_g$	Σ^+ (2), Σ^-, Π (3), Δ (2), Φ
$D_g + D_g$ or $D_u + D_u$	Σ^+ (3), Σ^- (2), Π (4), Δ (3), Φ (2), Γ
$D_g + D_u$	Σ^+ (2), Σ^- (3), Π (4), Δ (3), Φ (2), Γ

Table 2.2. Multiplicities of molecular states obtained by combining atomic states

Separated atoms	Molecule
Singlet + singlet	Singlet
Singlet + doublet	Doublet
Singlet + triplet	Triplet
Doublet + doublet	Singlet, triplet
Doublet + triplet	Doublet, quartet
Triplet + triplet	Singlet, triplet, quintet

Table 2.3. Molecular states obtained from two like atoms in identical electronic states

States of separated atoms	Molecular states
$^1S + {}^1S$	$^1\Sigma_g^+$
$^2S + {}^2S$	$^1\Sigma_g^+$, $^3\Sigma_u^+$
$^3S + {}^3S$	$^1\Sigma_g^+$, $^3\Sigma_u^+$, $^5\Sigma_g^+$
$^1P + {}^1P$	$^1\Sigma_g^+$ (2), $^1\Sigma_u^-$, $^1\Pi_g$, $^1\Pi_u$, $^1\Delta_g$
$^2P + {}^2P$	$^1\Sigma_g^+$ (2), $^1\Sigma_u^-$, $^1\Pi_g$, $^1\Pi_u$, $^1\Delta_g$, $^3\Sigma_u^+$ (2), $^3\Sigma_g^-$, $^3\Pi_g$, $^3\Pi_u$, $^3\Delta_u$
$^3P + {}^3P$	Singlet and triplet states as for $^2P + {}^2P$ plus $^5\Sigma_g^+$ (2), $^5\Sigma_u^-$, $^5\Pi_g$, $^5\Pi_u$, $^5\Delta_g$

combination of ^1S_g and ^1D_g, from table 2.1 and 2.2 we deduce that we will get $^1\Sigma^+$, $^1\Pi$ and $^1\Delta$ states for unlike atoms. For the case of like atoms we get $^1\Sigma_g^+$, $^1\Sigma_u^+$, $^1\Pi_g$, $^1\Pi_u$ and $^1\Delta_g$, $^1\Delta_u$ states.

2.6. The Term Manifold of Electronic States from the Molecular Orbital Configuration

In the previous section we discussed the use of the Wigner–Witmer rules for the determination of the manifold of electronic states correlating with a particular dissociation asymptote. It is also important to be able to deduce the states arising from a particular molecular orbital occupancy. Open shell configurations result in several electronic states. We again assume that spin–orbit coupling is relatively unimportant. The cases of non-equivalent electrons and equivalent electrons should be considered separately. By equivalent electrons we mean those in the same orbital or set of degenerate orbitals, such as the two electrons in the π_g orbital of O_2. Non-equivalent electrons are those in different orbitals.

We take the case of non-equivalent electrons first. Each molecular orbital can be characterized by a quantum number λ_i which gives the component of the orbital angular momentum along the internuclear axis. For a σ-orbital λ is zero, for a π-orbital $\lambda = \pm 1$ and for a δ-orbital $\lambda = \pm 2$. The resultant orbital angular momentum Λ about the internuclear axis is given by

$$\Lambda = \sum_i \lambda_i. \tag{2.41}$$

Similarly the resultant spin angular momentum S is given by

$$S = \sum_i s_i, \tag{2.42}$$

where s_i are the components of the spin angular momentum along the internuclear axis. Since we are concerned with non-equivalent electrons, the Pauli principle is automatically satisfied and any allowed value of S can be combined with any allowed value of Λ. Thus if we consider two electrons in different σ-orbitals, $\lambda_1 = 0$, $s_1 = \frac{1}{2}$; $\lambda_2 = 0$, $s_2 = \frac{1}{2}$ and $\Lambda = 0$ and $S = 0, 1$. Thus we have two states $^1\Sigma^+$ and $^3\Sigma^+$. If we have two π-electrons $\lambda_1 = \pm 1$, $s_1 = \frac{1}{2}$; $\lambda_2 = \pm 1$, $s_2 = \frac{1}{2}$; $|\Lambda| = 2, 0$ and $S = 1, 0$. The value $\Lambda = 0$ arises in two ways for $\lambda_1 = 1$, $\lambda_2 = -1$ and $\lambda_1 = -1$, $\lambda_2 = 1$ and consideration of the molecular wavefunctions shows that this gives rise to Σ^+ and Σ^- states. Thus we obtain six states $^3\Delta$, $^1\Delta$, $^3\Sigma^+$, $^1\Sigma^+$, $^3\Sigma^-$ and $^1\Sigma^-$. Table 2.4 lists the terms arising from various molecular orbital configurations.

If we have equivalent electrons, care must be taken not to violate the Pauli principle. The two electrons must differ in either λ_i or s_i. For the

Table 2.4. Electronic states for configurations of non-equivalent electrons

Electronic configuration	Molecular electronic states
σ	$^2\Sigma^+$
π	$^2\Pi$
$\sigma\sigma$	$^1\Sigma^+$, $^3\Sigma^+$
$\sigma\pi$	$^1\Pi$, $^3\Pi$
$\pi\pi$	$^1\Sigma^+$, $^3\Sigma^+$, $^1\Sigma^-$, $^3\Sigma^-$, $^1\Delta$, $^3\Delta$
$\sigma\sigma\sigma$	$^2\Sigma^+$, $^2\Sigma^+$, $^4\Sigma^+$
$\sigma\sigma\pi$	$^2\Pi$, $^2\Pi$, $^4\Pi$
$\sigma\pi\pi$	$^2\Sigma^+$ (2), $^4\Sigma^+$, $^2\Sigma^-$ (2), $^4\Sigma^-$, $^2\Delta$ (2), $^4\Delta$
$\pi\pi\pi$	$^2\Pi$ (6), $^4\Pi$ (3), $^2\Phi$ (2), $^4\Phi$

configuration π^2 which arises, for example, in the O_2 molecule where we consider only the electrons in the π_g (antibonding) orbital, $|\Lambda|$ can equal 2 or 0. The allowed occupancies of the orbitals are π_{+1}^2, π_{-1}^2 and $\pi_{+1}^1\pi_{-1}^1$. In the case of the configurations π_{+1}^2 and π_{-1}^2, the electrons must be singlet paired in order to satisfy the Pauli principle. This results in wavefunctions of the form

$$\pi_{+1}(1)\pi_{+1}(2)\frac{1}{\sqrt{2}}[\alpha(1)\beta(2)-\alpha(2)\beta(1)]. \qquad (2.43)$$

The resultant molecular orbital angular momentum, Λ, along the internuclear axis has the values $+2$, -2 for these two configurations which form the two components of a $^1\Delta$ state. Four permutations of the electrons exist for $\pi_{+1}^1\pi_{-1}^1$, namely $\pi_{+1}^1\alpha\pi_{-1}^1\alpha$, $\pi_{+1}^1\beta\pi_{-1}^1\beta$, $\pi_{+1}^1\alpha\pi_{-1}^1\beta$ and $\pi_{+1}^1\beta\pi_{-1}^1\alpha$. Only those combinations are allowed which satisfy the Pauli principle and have the correct symmetry. These considerations lead to a triplet of three wavefunctions which are of Σ^- symmetry

$$[\pi_{+1}(1)\pi_{-1}(2)-\pi_{+1}(2)\pi_{-1}(1)]\alpha(1)\alpha(2)$$

$$[\pi_{+1}(1)\pi_{-1}(2)-\pi_{+1}(2)\pi_{-1}(1)]\frac{1}{\sqrt{2}}[\alpha(1)\beta(2)+\alpha(2)\beta(1)] \qquad (2.44)$$

$$[\pi_{+1}(1)\pi_{-1}(2)-\pi_{+1}(2)\pi_{-1}(1)]\beta(1)\beta(2)$$

and one singlet wavefunction describing a $^1\Sigma^+$ state

$$[\pi_{+1}(1)\pi_{-1}(2)+\pi_{+1}(2)\pi_{-1}(1)]\frac{1}{\sqrt{2}}[\alpha(1)\beta(2)-\alpha(2)\beta(1)]. \qquad (2.45)$$

Table 2.5 gives the allowed terms for other configurations of equivalent electrons.

For details of terms arising from a configuration containing some equivalent and some non-equivalent electrons, the reader should consult Herzberg (1950).

Table 2.5. Electronic states for configurations of equivalent electrons

Electronic configuration	Molecular electronic states
σ^2	$^1\Sigma^+$
π^2	$^1\Sigma^+$, $^3\Sigma^-$, $^1\Delta$
π^3	$^2\Pi$
π^4	$^1\Sigma^+$

2.7. The Non-crossing Rule

In a given molecule one frequently has two potential curves of the same spin and symmetry correlating with different dissociation limits. For example in the series of molecules BeH, BH^+, MgH, AlH^+, $^2\Sigma$ states result from the interaction of $X(^1S_g) + H(^2S_g)$ (X = Be, B^+, Mg, Al^+) and $X(^3P_u) + H(^2S_g)$. On qualitative grounds one might expect the interaction between the closed shell of X in the ^1S state and the hydrogen atom to be repulsive whereas the interaction with the excited state ^3P might be expected to be attractive. It is therefore conceivable that the two curves would cross as indicated in figure 2.4. However, two states having the same spin multiplicity and the same value of Λ interact repulsively in such a way that the crossing is avoided. This leads to the non-crossing rule which says that in a diatomic molecule, for an infinitely slow change of internuclear distance, which is what was earlier described as the adiabatic Born–Oppenheimer separation, two electronic states of the same spin and symmetry cannot cross each other. This rule thus applies to adiabatic potential curves and may be rephrased to state that the potential curves of two electronic states of the same species (i.e., spin and symmetry) cannot cross each other. This is indicated qualitatively in figure 2.4 by dashed lines. The curves which would have intersected are often known as diabatic curves.

In the case of the hydrides BeH, BH^+, MgH and AlH^+, the result is two bound curves. The excited state has an abnormally long bond length and the ground state has a relatively shallow well as shown in figure 2.5 for BH^+ (see, for example, Guest and Hirst 1981a, 1981b). A similar situation arises in the case where an ionic curve is expected to cross a covalent curve. For example in NaCl the asymptote for $Na^+(^1S_g) + Cl^-(^1S_g)$ lies at an energy of 1.142 eV (given by the difference between the ionization potential of Na and the electron affinity of Cl) above that for $Na(^2S_g) + Cl(^2P_u)$. The ionic curve is strongly attractive because of the Coulomb attraction between the two ions whereas the covalent curve is initially almost perfectly horizontal. The two curves would be expected to intersect at an internuclear separation of about 10 Å. However there is an avoided intersection as shown in figure 2.6.

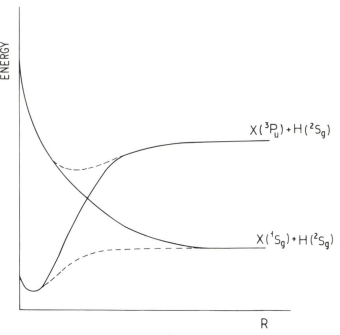

ENERGY

$X(^3P_u) + H(^2S_g)$

$X(^1S_g) + H(^2S_g)$

R

Figure 2.4. An avoided crossing between two $^2\Sigma^+$ states. Broken curves indicate the potential curves resulting from an avoided crossing between full curves.

In several cases the existence of an avoided intersection leads to a potential curve which has a local maximum as shown in figure 2.7. In such cases there is the possibility of the existence of vibrational levels which are higher in energy than the dissociation limit. Such levels are known as quasi-bound states. An example is the $C^2\Sigma^+$ state of CH which shows a local maximum in the region of $R = 1.7\,\text{Å}$. This is thought to arise from an avoided crossing between a covalent state correlating with $C(^1D) + H(^2S)$ and an attractive ionic state correlating with $C^-(^2P) + H^+(^1S)$. The existence of this potential maximum was deduced from the observation that the $v = 4$ vibrational level for CD lies slightly above the dissociation limit $C(^1D) + D(^2S)$ (Herzberg and Johns 1969) and has been confirmed by *ab initio* calculations (Lie *et al.* 1973).

It is also possible for avoided crossings to result in potential curves which have two minima. The potential curves for the excited $EF\,^1\Sigma_g^+$ and $GK\,^1\Sigma_g^+$ states of H_2 are examples of this. In the term symbols for these states the first letter (E or G) refers to the minimum at shorter R values. The wavefunction for the EF state is predominantly the configuration $1s\sigma_g 2s\sigma_g$ in the region of the inner minimum. The maximum occurs because of an avoided intersection with the configuration $2p\sigma_u^2$ and a further crossing with

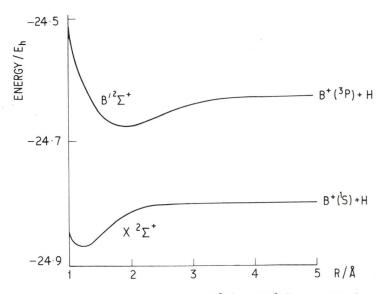

Figure 2.5. Potential curves for the $X\,^2\Sigma^+$ and $B'\,^2\Sigma^+$ states of BH$^+$.

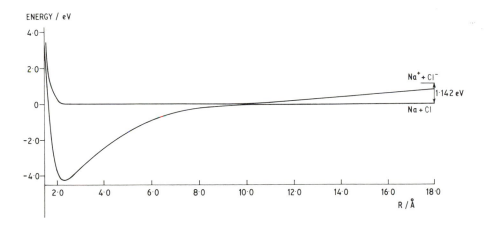

Figure 2.6. Potential curves for NaCl showing avoided crossing between covalent and ionic curves.

a curve corresponding to $H^+ + H^-$ leads to the outer minimum. In the case of the *GK* state, the inner minimum corresponds to the configuration $1s\sigma_g 3d\sigma_g$. As in the *EF* state the maximum is due to an avoided crossing with the configuration $2p\sigma_u^2$. The outer minimum is given by an avoided crossing with a curve corresponding to $H(1s) + H(n = 2)$. These curves are illustrated in figure 2.8.

Potential Energy Surfaces

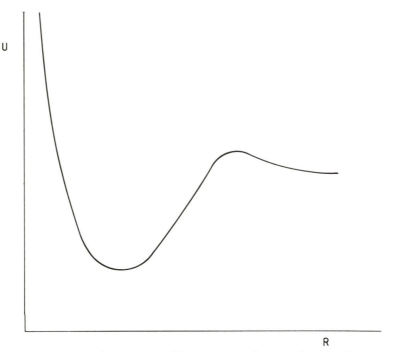

Figure 2.7. A diatomic potential energy curve showing a local maximum.

The occurrence of an avoided crossing can be seen from the following argument. Let us suppose that we know all the electronic wavefunctions except two. If ϕ_1 and ϕ_2 are two mutually orthogonal functions which are also orthogonal to all the electronic wavefunctions for the other states, then the wavefunctions ψ_1 and ψ_2 for these two states can be written as linear combinations of ϕ_1 and ϕ_2 thus

$$\psi_1 = c_{11}\phi_1 + c_{12}\phi_2$$
$$\psi_2 = c_{21}\phi_1 + c_{22}\phi_2. \qquad (2.46)$$

The energies E of the states and the wavefunctions are obtained from a linear variational treatment yielding the secular determinant

$$\begin{vmatrix} H_{11} - E & H_{12} \\ H_{12} & H_{22} - E \end{vmatrix} = 0 \qquad (2.47)$$

where H_{ij} represents the integral

$$\int \phi_i H \phi_j \, d\tau$$

in which H is the Hamiltonian operator. This equation is readily solved for E to give two roots

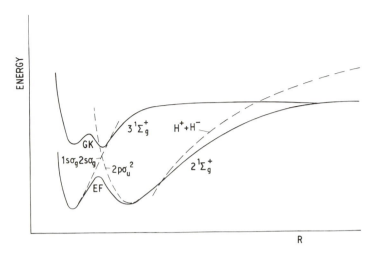

Figure 2.8. Potential curves for the *EF* and *GK* states of H_2.

$$E = \tfrac{1}{2}(H_{11} + H_{22}) \pm \tfrac{1}{2}[(H_{11} - H_{22})^2 + 4H_{12}^2]^{\frac{1}{2}}. \qquad (2.48)$$

The potential curves can cross only if the two energies are equal. For this to be the case we require $H_{11} = H_{22}$ and simultaneously $H_{12} = 0$. The integral H_{12} will only equal zero for all values of R if ϕ_1 and ϕ_2 belong to different species, provided that spin–orbit coupling terms are absent from the Hamiltonian operator. If there is a value of R for which $H_{11} = H_{22}$ the potential curves can cross. However, if ϕ_1 and ϕ_2 are of the same species, H_{12} will, in general, be non-zero and even if $H_{11} = H_{22}$ the potential curves will not cross. The form of equation 2.48 indicates that the two curves repel each other. At some distance from the intersection the functions ϕ_1 and ϕ_2 (for that particular value of R) will be reasonable representations of the wavefunctions for the two states. If to the left of the intersection ϕ_1 describes the lower state and ϕ_2 the upper, we expect that to the right of the intersection ϕ_1 describes the upper state and ϕ_2 the lower state (see figure 2.9). In the region of the avoided crossing the wavefunction for the lower state changes from ϕ_1 to ϕ_2 as R increases. However, care is sometimes necessary in interpreting the results of *ab initio* configuration interaction calculations and it is preferable to supplement wavefunction calculations by additional calculations such as the dipole moment function.

The functions ψ_1 and ψ_2 are the wavefunctions for the adiabatic potential curves and ϕ_1 and ϕ_2 are diabatic wavefunctions. Diabatic (intersecting) potential curves are defined by the variation of the integrals H_{11} and H_{22} with R. The concept of diabatic states is not as well defined as the

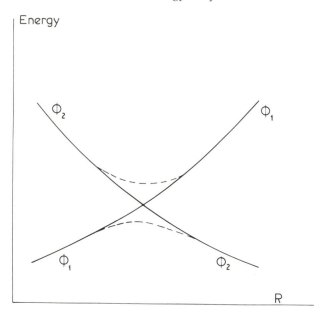

Figure 2.9. An avoided crossing between two diabatic functions ϕ_1 and ϕ_2.

adiabatic case (see Baede (1975), Lichten (1967) and Smith (1969) for a fuller discussion).

The Born–Oppenheimer separation is not valid in the region of narrowly avoided intersections, and coupling terms involving the nuclear kinetic energy operator can give rise to transitions between adiabatic states. This will be discussed further in chapter 6.

2.8. Intersection Between Potential Curves for Different Electronic States

2.8.1. *Perturbations in the Rotational Energy Levels*

We have so far assumed that the Born–Oppenheimer separation is valid and that the Hamiltonian operator does not contain any terms involving spin. At this level of approximation, integrals

$$H_{ij} = \int \phi_i H \phi_j \; d\tau$$

are zero if ϕ_i and ϕ_j belong to states differing in spin or symmetry. Thus there is no interaction between the two states and the potential curves are allowed to cross. However if spin–orbit coupling and the coupling between the electronic orbital angular momentum and the total angular momentum are included in the Hamiltonian, the integrals are no longer necessarily zero and although the potential curves cross, the energy levels associated with the two states may interact. If two potential curves for bound states cross, any interaction will be manifested in perturbations in the spectrum. Kronig showed that for the integral H_{ij} to be non-zero, several conditions have to be satisfied. We are now concerned with the total wavefunction for a particular vibration–rotation level of a given electronic state and not just with the electronic wavefunction. The conditions are as follows:

1. The quantum number J for the total angular momentum must be the same in both states, i.e. $\Delta J = 0$.

2. Both states should be of the same spin multiplicity, i.e. $\Delta S = 0$.

3. The difference in $|\Lambda|$ values (for the electronic orbital angular momentum) for the states must be 0 or ± 1, i.e. $\Delta |\Lambda| = 0, \pm 1$.

4. Positive-parity rotational levels cannot interact with negative-parity rotational levels.

5. For molecules with identical nuclei, s states cannot combine with a states.

The second and third conditions are not completely rigorous. The perturbations are referred to as homogeneous if $\Delta \Lambda = 0$ and heterogeneous if $\Delta |\Lambda| = \pm 1$. Since we are interested here in the case where two potential curves intersect, we shall be concerned with heterogeneous perturbations. At the level where the $\Delta \Lambda$ criterion is valid, one only expects perturbations between Σ and Π states, between Π and Δ states etc. Σ^+ and Σ^- states do not interact because for a given value of J one level will be $+$ and the other $-$. Also a Π_g state cannot interact with a Π_u state.

Two levels will interact only if they have comparable energies and in addition to the above criteria there also has to be an appreciable overlap between the vibrational wavefunctions for the two states. This is analogous to the Franck–Condon principle. Thus perturbations will be significant only in the region of intersection of the two potential curves. Even though levels 1 and 2 in figure 2.10 have approximately the same energy and satisfy the Kronig criteria, there will be negligible interaction between them because of the very small overlap between the two vibrational wavefunctions. However levels 3 and 4, provided that the Kronig criteria are satisfied, should interact strongly because there will be appreciable overlap between the two vibrational wavefunctions.

The result of the interaction is given by an expression of the form of equation 2.48 and the two interacting levels are pushed apart. Some interaction will occur for levels adjacent to the two nearly coincident levels ($J = 4$ in figure 2.11) but will decrease rapidly as one moves to higher or lower J values. Thus there will be a very noticeable irregularity in rotational spacings as one passes through the J values in the region where the levels would

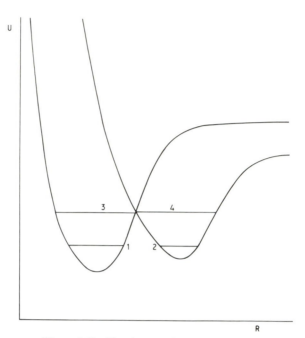

Figure 2.10. Two intersecting potential curves.

be nearly coincident in the absence of the perturbation. From an analysis of the perturbations it is often possible to deduce some spectroscopic constants for the perturbing state, even though transitions to that state are not observed.

2.8.2. Predissociation

In the previous section we were concerned with perturbations in the rotational structure arising from the intersection of two bound states. Intersections also occur between bound and repulsive potential curves, as illustrated in figure 2.12. There is now the possibility of a radiationless transition occurring from the bound state to the unbound state resulting in dissociation of the molecule. This is known as predissociation. However, it does not occur for every intersection of a bound potential curve with a repulsive curve, as predissociation can only occur when the Kronig criteria are satisfied. The energy levels for the repulsive potential form a continuum and the 'vibrational' motion is not quantized, but one can still talk in terms of the rotational motion being quantized, and levels in the continuum can be characterized by a value for J and by $+$, $-$ and s, a character. For any given energy

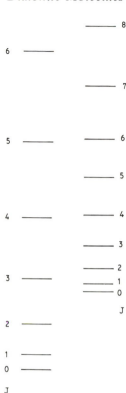

Figure 2.11. Schematic diagram of rotational levels for two intersecting states.

level in the bound state it is always possible to satisfy the $\Delta J = 0$ criterion and provided that the other conditions are satisfied predissociation can occur. The vibrational criterion discussed in the previous section is also applicable to the case of predissociation. There has to be an appreciable overlap between the vibrational wavefunction for the bound state and the wavefunction for the level in the continuum. This will hold only in the region of the intersection of the two curves.

The occurrence of predissociation in a given molecule is indicated by diffuseness in the rotational structure in the absorption spectrum or sudden loss of intensity in an emission spectrum which starts from or terminates on the predissociating state. In the absence of other effects, the width of a spectral line is inversely proportional to the lifetime of the molecule in the state in question. This is a consequence of the Heisenberg uncertainty principle. If radiationless transitions are taking place from a given vibration–rotation level of the bound state to the dissociative state, the lifetime of that level is reduced and the line-width of transitions to that level is increased. If we go to higher levels in the bound state, the vibrational overlap will be

Potential Energy Surfaces

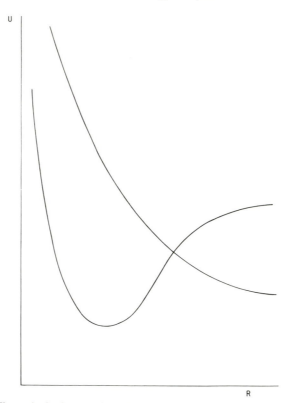

Figure 2.12. Intersection of a bound state with a repulsive state.

negligible, predissociation will not occur and the rotational structure becomes sharp again.

Predissociation can also occur when two bound states intersect as shown in figure 2.13. Transitions can occur from the ground state X to the excited state A which intersects another bound state B having a dissociation limit lower than that of A. If excitation occurs to a vibration–rotation level of A which is comparable in energy to but slightly greater than the energy at the intersection of A and B (as drawn in figure 2.13), radiationless transitions from A to B are very likely if the Kronig criteria are satisfied. However, once the transition to state B has occurred, the molecule may have energy in excess of its dissociation energy and will decompose within one vibrational period.

There is another type of predissociation which can occur for diatomic molecules. This is known as predissociation by rotation. In the radial Schrödinger equation for the vibration–rotation levels of a diatomic molecule (equation 2.6) there is a term $J(J+1)/R^2$ arising from the potential energy associated with the centrifugal force due to rotational angular momentum.

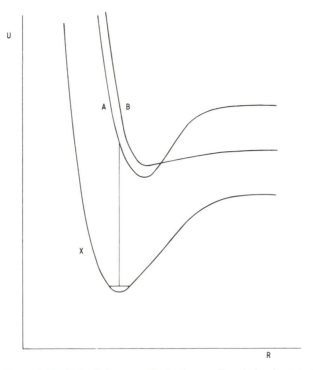

Figure 2.13. Potential curves illustrating predissociation in state A.

The equation can be rewritten

$$\frac{1}{R^2}\frac{d}{dR}\left(R^2\frac{d\psi}{dR}\right) + \frac{8\pi^2\mu}{h^2}[E - U_{\text{eff}}(R)]\psi = 0 \qquad (2.49)$$

where $U_{\text{eff}}(R)$ is an effective potential energy given by

$$U_{\text{eff}}(R) = U(R) + \frac{h^2}{8\pi^2\mu R^2}J(J+1). \qquad (2.50)$$

Provided that J is not too large, the result is a potential curve having a local maximum as shown in figure 2.14. Thus there is the possibility that there will be some vibration–rotation levels having an energy above that of the dissociation limit. If there is an appreciable probability of quantum mechanical tunnelling through the barrier, dissociation will occur and the corresponding rotational lines in the spectrum will be diffuse.

From the line widths or lifetimes of the levels showing predissociation it is possible to deduce the potential curve for the repulsive state in the region in which predissociation occurs. This is discussed in detail by Child (1974) in a comprehensive review of the dynamics of predissociation.

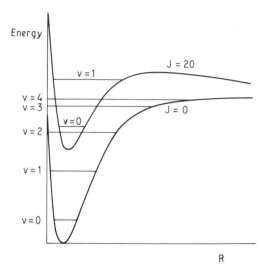

Figure 2.14. Effective potential curves for states with $J = 0$ and $J = 20$.

Suggestions for Further Reading

G. Herzberg (1950) *Molecular Spectra and Molecular Structure,* Vol. I. *Spectra of Diatomic Molecules,* 2nd ed.,Van Nostrand, Princeton.

J. M. Hollas (1982) *High Resolution Spectroscopy,* Butterworths, London.

G. W. King (1964) *Spectroscopy and Molecular Structure,* Holt, Rinehart and Winston, New York.

References

A. P. M. Baede (1975) in *Molecular Scattering: Physical and Chemical Applications,* ed. K. P. Lawley, p. 463, Wiley, London.

J. K. Cashion (1963) *J. Chem. Phys.,* **39**, 1872.

M. S. Child (1974) in *Molecular Spectroscopy* Vol. 2, ed. R. F. Barrow, D. A. Long and D. J. Millen, p. 466, The Chemical Society, London.

J. W. Cooley (1961) *Math. Computation,* **15**, 363.

W. Gordy and R. I. Cook (1970) *Microwave Molecular Spectra, Chemical Applications of Spectroscopy, part II,* ed. W. Weser, Interscience, New York.

M. F. Guest and D. M. Hirst (1981a) *Chem. Phys. Lett.,* **80**, 131.

M. F. Guest and D. M. Hirst (1981b) *Chem. Phys. Lett.,* **84**, 167.

G. Herzberg (1950) *Molecular Spectra and Molecular Structure,* Vol I. *Spectra of Diatomic Molecules,* 2nd ed., Van Nostrand, Princeton.

G. Herzberg and J. W. C. Johns (1969) *Astrophys. J.,* **158**, 399.

K. P. Huber and G. Herzberg (1979) *Molecular Spectra and Molecular Structure,* Vol. IV. *Constants of Diatomic Molecules,* Van Nostrand Reinhold, New York.

P. Huxley and J. N. Murrell (1983) *J. Chem. Soc. Farad. II,* **79**, 323.

R. J. LeRoy (1979) *Chemical Physics Research Report CP–110R,* University of Waterloo, Canada.

W. P. Lichten (1967) *Phys. Rev.,* **164**, 131.

G. C. Lie, J. Hinze and B. Liu (1973) *J. Chem. Phys.,* **59**, 1872.

E. A. Mason and L. Monchick (1967) in *Intermolecular Forces,* ed. J. O. Hirschfelder, p. 329, Wiley–Interscience, New York.

I. M. Mills (1974) in *Theoretical Chemistry,* Vol. 1, ed. R. N. Dixon, p. 110, The Chemical Society, London.

J. N. Murrell and K. S. Sorbie (1974) *J. Chem. Soc. Farad. II,* **70**, 1552.

L. Pauling and E. B. Wilson (1935) *Introduction to Quantum Mechanics,* McGraw-Hill, New York.

F. T. Smith (1969) *Phys. Rev.,* **179**, 111.

J. T. Vanderslice, E. A. Mason, W. G. Maisch and E. R. Lippincott (1959) *J. Mol. Spect.,* **3**, 17.

CHAPTER 3

triatomic molecules

3.1. Potential Hypersurfaces for Triatomic Molecules

Whereas for diatomic molecules the potential energy function is simply a function of the internuclear distance, for a polyatomic molecule containing N atoms, the potential energy function depends on $3N-6$ internuclear distances. Thus for a triatomic species the potential energy function depends on three variables and it is not possible to make a full graphical representation of this. In portraying polyatomic potential surfaces one has to impose some constraints on the variables. For linear molecules or for reactive processes for which a collinear approach is most likely, a useful representation of the surface is for collinear geometries. For the molecule ABC the potential energy is then simply a function of R_{AB} and R_{BC}. The most informative graphical representation is in the form of a contour diagram in which the axes represent the internuclear distances R_{AB} and R_{BC} and the contours join up geometries of the same energy.

A collinear potential surface for the reaction

$$F + H_2 \rightarrow FH + H \qquad (3.1)$$

is represented in figure 3.1. Points for large R_{FH} with R_{HH} equal to the equilibrium bond length of H_2 represent the reactants at large separations where there is no interaction between them. A cut through the surface in this region, for constant R_{FH}, is simply the potential energy curve for H_2. Similarly the products are represented by points for which R_{FH} is equal to the equilibrium bond length of HF and the value of R_{HH} is large. A cut here, for constant R_{HH}, is the potential energy curve for HF. The plateau region for large values of R_{FH} and R_{HH} represents the dissociation limit for separated atoms $F + H + H$.

If we start at a point A corresponding to reactants at large separation and let R_{FH} decrease, the energy steadily increases if R_{HH} is more or less constant. It is as though we were travelling up a valley. However, motion in a perpendicular direction results in a rapid increase in energy. If we continue along the bottom of the valley, following the minimum energy pathway, the energy continues to increase until we reach the point B, which is a saddle

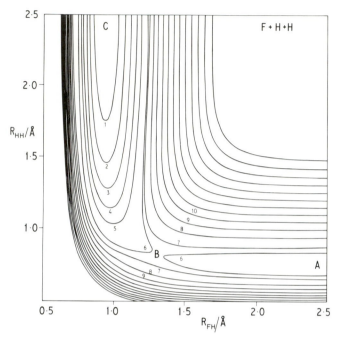

Figure 3.1. Collinear potential energy surface for FHH.

point or col. Motion forwards or backwards along the minimum energy path results in a decrease in energy but motion in a perpendicular direction results in a very large increase in energy, particularly in the direction for which R_{FH} and R_{HH} are both decreasing. The saddle point B corresponds to the top of the activation barrier in simple collision theory or in transition state theory. However there is no clear relationship between the height of the saddle point relative to reactants and the empirical activation energy obtained from the Arrhenius equation. The valley leading up to the saddle point is known as the entrance channel. Continuation along the minimum energy path from the saddle point results in a steady decrease in energy along the exit channel until one reaches products at large separation (point C). The minimum energy pathway is known as the reaction coordinate but this is also difficult to define precisely in a unique way. The dynamics of a reaction are very dependent on the position of the saddle point. If this occurs in the entrance channel, for example, translational energy is more effective than vibrational energy in surmounting the barrier and the product is vibrationally excited. The differences in dynamics will be discussed in detail in chapter 6.

A collinear potential energy surface can also be portrayed by a two-dimensional isometric projection of the three-dimensional surface. Such diagrams can be obtained readily by the use of computer graphics packages and although they do sometimes make it easier to visualize some features of the surface, they are generally less useful than a conventional contour diagram because it is not easy to extract numerical data from them.

For other reactive systems, the perpendicular approach of A to BC may be energetically preferred. In such cases the most useful representation of the potential surface is in terms of the BC distance and the perpendicular distance of A from some appropriate point between B and C (see figure 3.2).

Figure 3.2. Definition of parameters for C_{2v} geometries.

If BC is a homonuclear diatomic molecule, the mid-point of BC would be an appropriate point. An example of a potential energy surface portrayed in this way is given in figure 3.3 for the case of $O(^1D) + H_2$, with the HH distance being represented along the x-axis.

Interesting phenomena occur when two potential surfaces are in close proximity to each other. It is difficult to represent this by a single contour diagram and one really has to superpose one contour diagram on top of the other. The resulting composite diagram may be very confused and the situation is more easily visualized in terms of a graph showing cuts through the relevant surfaces in which one parameter is fixed and the other is allowed to vary. Figure 3.4 shows cuts through the potential surfaces for the reaction of N^+ with H_2 for perpendicular geometries with R_{HH} fixed at a value of $2\,a_0$ ($1\,a_0 = 5.291\,7706 \times 10^{-11}\,m$) (Hirst 1978). The interesting feature, for the interpretation of the dynamics, is the intersection between the $^3A_2$ and $^3B_1$ surfaces. If the geometry is distorted slightly from perpendicular, the symmetry is reduced from C_{2v} to C_s and the surfaces will both be of the same symmetry ($^3A''$). By analogy with the diatomic case discussed in section 2.7, the crossing is avoided. Avoided crossings in triatomic surfaces will be discussed in section 3.3.

We turn now to a discussion of potential surfaces for bound triatomic molecules. Figure 3.5 illustrates the collinear potential surface for the molecule HCN which dissociates to $H(^2S) + CN(^2\Sigma^+)$ and $HC(^4\Sigma^-) + N(^4S)$. The well in the surface represents the potential minimum corresponding to the equilibrium configuration of HCN. A cut through the surface for a large constant value of R_{HC} gives a representation of the diatomic potential curve

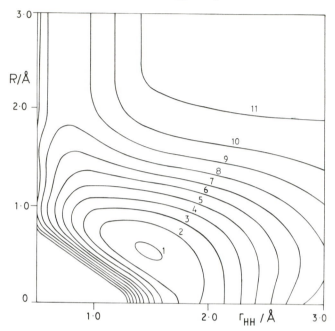

Figure 3.3. Potential energy surface for C_{2v} geometries of $O(^1D)+H_2$. Contour 1 = -10.25 eV. Contours drawn at intervals of 1 eV. (Reproduced from SERC Potential Energy Data Base.)

for $CN(^2\Sigma^+)$ and the corresponding cut for a constant large value of R_{CN} represents the potential curve for $HC(^4\Sigma^-)$. The plateau region for large R_{HC} and large R_{CN} represents the limit of complete dissociation to $H+C+N$. The two vibrational stretching modes for HCN, usually designated ν_1 and ν_3, are essentially CN and CH vibrations respectively. These modes, for small distortions from equilibrium, are indicated in the diagram. Although a diagram of the form of figure 3.5 is very useful for consideration of linear motion in HCN, it fails to illustrate some very important and interesting features of this molecule because it can portray only linear structures of the molecule with the atoms in the order H...C...N. The isomer HNC exists as a stable entity and the isomerization process

$$HCN \rightarrow HNC \tag{3.2}$$

is of considerable interest. In order to discuss this process we need a different representation of the HCN potential energy surface. One way of doing this takes the centre of the CN bond as the origin and represents the position of the hydrogen atom relative to this by the coordinates (x_H, y_H) (see figure 3.6). Energy contours are then plotted for motion of the

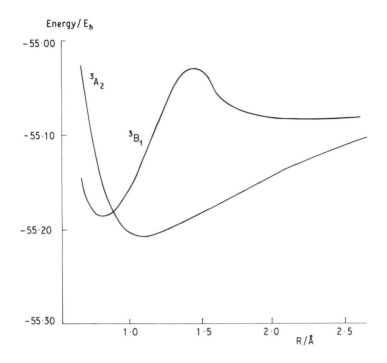

Figure 3.4. Cuts through $^3B_1$ and $^3A_2$ potential energy surfaces for NH_2^+ for $r_{HH} = 2 a_0$.

hydrogen atom around the CN fragment. In order to fully represent the changes in energy that occur as the hydrogen atom is moved, the CN internuclear distance should be optimized for each position of the hydrogen atom. Figure 3.7 illustrates the HCN $\rightarrow$ HNC potential surface portrayed in this way. Two potential minima A and B are apparent, corresponding to the equilibrium geometries for the two isomers HCN and HNC and the minimum energy pathway between the two isomers can be seen. A saddle point occurs at the configuration marked S.

A third form of contour diagram is useful for X_3 molecules for which the potential surface has several symmetrically equivalent minima. An example is the molecule O_3 for which the potential surface will have three equivalent minima. Rather than portraying the potential energy function in terms of internuclear distances R_{AB}, or deviations ρ_{AB} from a reference structure with three equal internuclear distances R_{AB}^0, $\rho_{AB} = R_{AB} - R_{AB}^0$, a set of symmetry adapted displacement coordinates S_i is defined in terms of ρ_{AB} by the transformation

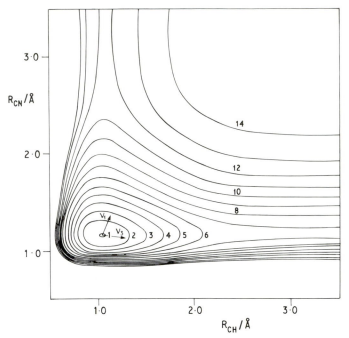

Figure 3.5. Collinear potential energy surface for HCN. Contour $1 = -13.75\,\text{eV}$. Contours drawn at intervals of $1\,\text{eV}$. (Reproduced from SERC Potential Energy Data Base.)

$$\begin{pmatrix} S_1 \\ S_2 \\ S_3 \end{pmatrix} = \begin{pmatrix} \dfrac{1}{\sqrt{3}} & \dfrac{1}{\sqrt{3}} & \dfrac{1}{\sqrt{3}} \\ \dfrac{1}{\sqrt{2}} & 0 & -\dfrac{1}{\sqrt{2}} \\ -\dfrac{1}{\sqrt{6}} & \dfrac{2}{\sqrt{6}} & -\dfrac{1}{\sqrt{6}} \end{pmatrix} \begin{pmatrix} \rho_{AB} \\ \rho_{BC} \\ \rho_{AC} \end{pmatrix} \qquad (3.3)$$

The coordinate S_1 transforms as the totally symmetric representation of the point group D_{3h}, whereas S_2 and S_3 transform as the E representation. Taking S_1 as a constant fixes the value of the perimeter of the triangle of the three atoms and the potential function can then be represented as a function of S_2 and S_3. The diagram is constrained to lie within an equilateral triangle with sides of length $(\tfrac{3}{2})^{\frac{1}{2}} S_1$ (figure 3.8). The sides of the triangle correspond to linear configurations. The symmetrical D_{3h} configuration is at the centroid of the triangle and C_{2v} configurations lie along perpendiculars from an apex to the opposite side (see Murrell 1980). An example of such a diagram is given in figure 3.8 for the ozone molecule. The three equivalent minima which correspond to the C_{2v} equilibrium structure of ozone are clearly visible. There is also a fourth minimum corresponding to a metastable state in

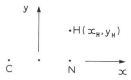

Figure 3.6. Geometrical parameters for HCN → HNC potential energy surface.

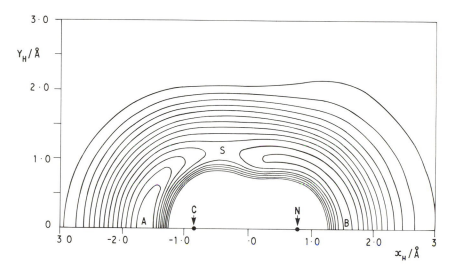

Figure 3.7. Potential energy surface for HCN → HNC isomerization. Contours drawn at intervals of $0 \cdot 4$ eV. (Reproduced from SERC Potential Energy Surface Data Base.)

which the geometry is an equilateral triangle.

As in the case of potential surfaces for reactive systems, it is often convenient to use a graph to illustrate some particular section of the potential surface as a function of one single coordinate only, such as the variation in energy on bending the molecule but with the restriction of fixed bond lengths.

3.2. Correlation Diagrams

In section 2.5 we discussed the use of the Wigner–Witmer rules to derive the manifold of electronic states of a diatomic molecule resulting from two atomic fragments. Here we consider the generalization of these ideas to triatomic species. The methods are also applicable to molecules containing four or more atoms and details can be found in Herzberg (1966).

Potential Energy Surfaces

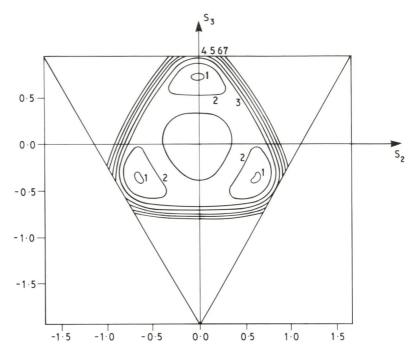

Figure 3.8. Potential energy surface for O_3 for a perimeter of 4.7325 Å. Contour 1 = −6.25 eV. Contours drawn at intervals of 1 eV. (Reproduced from SERC Potential Energy Surface Data Base.)

3.2.1. Linear Molecules

As in the case of diatomic molecules, unsymmetrical molecules and symmetrical molecules will be discussed separately. For unsymmetrical molecules, one approach is to start with the separated atoms and then obtain all the possible values of the components of orbital angular momentum (Λ) and spin angular momentum (S) along the internuclear axis. Suppose, for example, that we are interested in the states of $[\text{B—H—H}]^+$ arising from the excited state of $B^+(^3P)$ and two hydrogen atoms in the ground state (^2S). The L quantum numbers are equal to 1 for the boron atom and to 0 for the hydrogen atoms, respectively. Generalizing the discussion in section 2.5 to three atoms we get

$$\Lambda = M_A + M_B + M_C. \tag{3.4}$$

The M quantum number for the boron ion can have values 1, 0, and −1, so the allowed resultant values of Λ are 1, 0 and −1. For the spins we have to form a partial resultant from two of them and then combine that with the third spin. From the two hydrogen atoms H_1 and H_2 we obtain, from equation 2.40, $S_{H_1H_2} = 1, 0$. We then consider the combination of the boron spin angular momentum $S_B = 1$ with each of these two values of $S_{H_1H_2}$ to give the resultant values of 2, 1, 0 and 1. We may combine any of the spin quantum numbers with either of the orbital quantum numbers 1 or 0 to deduce the existence of $^5\Sigma, ^5\Pi, ^3\Sigma, ^3\Pi, ^1\Sigma, ^1\Pi$, and $^3\Sigma, ^3\Pi$ states. Parity considerations lead to the conclusion that the Σ states are Σ^+. For three atoms the state is Σ^+ or Σ^- depending on whether the sum

$$L_A + L_B + L_C + \sum l_{iA} + \sum l_{iB} + \sum l_{iC} \tag{3.5}$$

is even or odd, where L_A, L_B and L_C are the resultant orbital momenta for atoms A, B and C and the sums $\sum l_i$ for each atom are over all electrons outside closed subshells. If at least two of the M_L values are non-zero, the resulting Σ states will occur in pairs Σ^+, Σ^- in which the members of each pair occur by reversal of the signs of all the M_L values.

However for polyatomic molecules we are usually more interested in the states correlating with two particular fragments rather than with the limit of complete dissociation to atoms. Thus in the $[BHH]^+$ example we would be interested in states correlating with $B^+(^3P) + H_2(^1\Sigma_g^+)$ and would not be particularly concerned with states correlating with the repulsive $^3\Sigma_u^+$ state of H_2. The allowed values of the resultant orbital angular momentum Λ are obtained by using the formula

$$\Lambda = M_{L_A} + M_{L_{BC}} \tag{3.6}$$

where M_{L_A} can take any of the allowed values of the angular momentum of the atomic species, $L_A, L_A - 1, \ldots, -L_A$ and $M_{L_{BC}} = \Lambda_{BC}$ is the orbital angular momentum quantum number for the diatomic fragment. Thus for $B^+(^3P) + H_2(^1\Sigma_g^+)$ we obtain $\Lambda = 1, 0, -1$ resulting in Σ and Π states. The resultant spin angular momentum is obtained by taking the allowed combinations of the spin angular momentum on the atom with that on the molecule. In this case the only allowed resultant is $S = 1$ and we have $^3\Sigma$ and $^3\Pi$ states which correlate with $B^+(^3P) + H_2(^1\Sigma_g^+)$. Table 3.1 contains details of the triatomic states resulting from combinations of atoms in S, P and D states with molecular Σ^+, Σ^- and Π states.

In the qualitative discussion of reaction dynamics, it is useful to construct a correlation diagram showing the correlation between reactant asymptotes and product asymptotes. Figure 3.9 shows such a diagram for states of $[BHH]^+$ arising from $B^+(^1S, ^3P) + H_2(^1\Sigma_g^+)$ and $B(^2S) + H_2^+(^2\Sigma_g^+)$ and resulting in $BH^+(X^2\Sigma^+, A^2\Pi, B'^2\Sigma^+) + H(^2S)$ and $BH(^1\Sigma^+) + H^+(^1S)$. Such diagrams are purely qualitative and tell us nothing about the shape of the potential surface, although it is often possible to make some deductions about the surfaces from a knowledge of the orbital occupancies. More detailed discussions of correlation diagrams are given by Donovan and Husain (1970) and Mahan (1971).

Table 3.1. Molecular states obtained by combining an atom with a diatomic (or linear) molecule

States of separated atom and molecule	Resulting molecular state
$S_g + \Sigma^+$ or $S_u + \Sigma^-$	Σ^+
$S_g + \Sigma^-$ or $S_u + \Sigma^+$	Σ^-
$S_g + \Pi$ or $S_u + \Pi$	Π
$P_g + \Sigma^+$ or $P_u + \Sigma^-$	Σ^-, Π
$P_g + \Sigma^-$ or $P_u + \Sigma^+$	Σ^+, Π
$P_g + \Pi$ or $P_u + \Pi$	$\Sigma^+, \Sigma^-, \Pi, \Delta$
$D_g + \Sigma^+$ or $D_u + \Sigma^-$	Σ^+, Π, Δ
$D_g + \Sigma^-$ or $D_u + \Sigma^+$	Σ^-, Π, Δ
$D_g + \Pi$ or $D_u + \Pi$	$\Sigma^+, \Sigma^-, \Pi, \Pi, \Delta, \Phi$

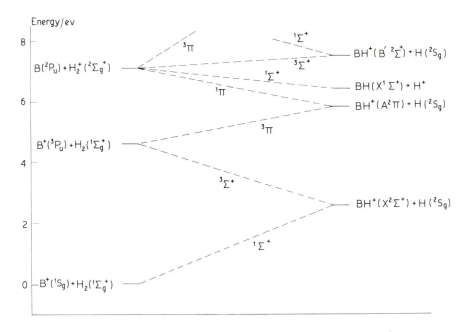

Figure 3.9. Correlation diagram for linear configurations of $[BHH]^+$.

Equation 3.6 can be generalized to the case of two different linear fragments for which the allowed values of M are M_{AB} ($= \Lambda_{AB}$) and M_{CD} ($= \Lambda_{CD}$) to give

$$\Lambda = M_{AB} + M_{CD}. \tag{3.7}$$

The allowed states arising from linear fragments in Σ or Π states are given in table 3.2.

Table 3.2. Molecular states obtained by combining two different linear fragments

States of separated groups	Resulting molecular states
$\Sigma^+ + \Sigma^+$ or $\Sigma^- + \Sigma^-$	Σ^+
$\Sigma^+ + \Sigma^-$	Σ^-
$\Sigma^+ + \Pi$ or $\Sigma^- + \Pi$	Π
$\Pi + \Pi$	$\Sigma^+, \Sigma^-, \Delta$

Table 3.3. Molecular states resulting from the combination of two like fragments in the same electronic state

States of separated groups	Resulting molecular states
$^1\Sigma^+ + {}^1\Sigma^+$ or $^1\Sigma^- + {}^1\Sigma^-$	$^1\Sigma_g^+$
$^2\Sigma^+ + {}^2\Sigma^+$ or $^2\Sigma^- + {}^2\Sigma^-$	$^1\Sigma_g^+, {}^3\Sigma_u^+$
$^3\Sigma^+ + {}^3\Sigma^+$ or $^3\Sigma^- + {}^3\Sigma^-$	$^1\Sigma_g^+, {}^3\Sigma_u^+, {}^5\Sigma_g^+$
$^1\Pi + {}^1\Pi$	$^1\Sigma_g^+, {}^1\Sigma_u^-, {}^1\Delta_g$
$^2\Pi + {}^2\Pi$	$^1\Sigma_g^+, {}^1\Sigma_u^-, {}^1\Delta_g, {}^3\Sigma_u^+, {}^3\Sigma_g^-, {}^3\Delta_u$
$^3\Pi + {}^3\Pi$	$^1\Sigma_g^+, {}^1\Sigma_u^-, {}^1\Delta_g, {}^3\Sigma_u^+, {}^3\Sigma_g^-, {}^3\Delta_u, {}^5\Sigma_g^+, {}^5\Sigma_u^-, {}^5\Delta_g$

For symmetrical molecules we consider three cases. When two identical fragments in different states are brought together we obtain the molecular states given in table 3.2. However, as in the diatomic case discussed in section 2.5, each state occurs twice, once with g symmetry and once with u symmetry. If the two fragments are in the same electronic state we only get the same number of states as in the case of unlike fragments. However it is necessary to determine whether the electronic state of the final molecule is g or u. Table 3.3 contains details of the states obtained from combinations of various fragments in Σ and Π states. The third case we consider is where a linear symmetrical molecule such as YXY is built up unsymmetrically from two fragments Y + XY. The table does not tell us whether the state is of g or u symmetry in the case of a symmetrical molecule. It may be possible in some cases to determine this by consideration of building up the molecule in a symmetric manner from atoms. If we consider the states of linear symmetric $[HBH]^+$ that can be obtained from $H(^2S_g) + BH^+(^2\Sigma^+)$ we obtain $^1\Sigma^+$ and $^3\Sigma^+$ states. If BH^+ is in a $^2\Pi$ state, we obtain $^1\Pi$ and $^3\Pi$ states of $[HBH]^+$. If we build the molecule symmetrically from $H(^2S_g) + B^+(^1S_g) + H(^2S_g)$, combination of the hydrogen atoms yields $^1\Sigma_g^+$ and $^3\Sigma_u^+$ states (from table 2.3). To combine $B^+(^1S_g)$ with these states of H_2 it is first necessary to resolve the atomic state ^1S_g in the field of $D_{\infty h}$ symmetry. This yields $^1\Sigma_g^+$. Thus we obtain $^1\Sigma_g^+$ and $^3\Sigma_u^+$ states for $[HBH]^+$. In the case of $B^+(^3P_u)$, resolution into $D_{\infty h}$ gives $^3\Sigma_u^+ + {}^3\Pi_u$ and the following states are obtained for $[HBH]^+$: $^3\Sigma_u^+$, $^3\Pi_u$, $^1\Sigma_g^+$, $^3\Sigma_g^+$, $^5\Sigma_g^+$, $^1\Pi_g$, $^3\Pi_g$, $^5\Pi_g$. Thus we can say unambiguously that the $^1\Pi$ state is $^1\Pi_g$ but the $^3\Pi$ state may be $^3\Pi_g$ or $^3\Pi_u$. For further details of the molecular states arising from various combinations of atomic fragments the reader is referred to Herzberg (1966).

3.2.2. Non-linear Molecules

If a non-linear polyatomic molecule is built up from separated atoms, a large number of molecular states will be formed in general. This is discussed in detail by Herzberg (1966) but we will restrict ourselves to a brief discussion of the states arising when a molecule is built up from two fragments. In order to obtain the states of the product molecule it is necessary to use group theory. If the molecule is built up from two unlike fragments X and Y, the first step is to resolve the states of X and Y into an appropriate common point group. The allowed states of the XY molecule are obtained by taking the direct product of the resolution from X and Y. Let us consider the formation of a bent $[HBH]^+$ species as $B^+(^3P_u)$ approaches $H_2(^1\Sigma_g^+)$ along a line bisecting the HH bond. The final molecule will be of C_{2v} symmetry and during its formation the system will have C_{2v} symmetry. Resolution of ^3P_u into C_{2v} yields $^3A_1 + ^3B_1 + ^3B_2$. Similarly resolution of $^1\Sigma_g^+$ into C_{2v} symmetry yields $^1A_1$. The states of $[HBH]^+$ are then obtained by taking the direct product of $(^3A_1 + ^3B_1 + ^3B_2)$ with $^1A_1$ which, of course, gives $^3A_1 + ^3B_1 + ^3B_2$. If the symmetry of the final molecule XY is the same as that of one of the fragments, say Y, it is only necessary to resolve the state of X into the point group of Y (and XY).

In some cases ambiguities do arise and it is not always possible to say exactly which state of the molecule will be obtained. For example, if we consider the formation of $[HBH]^+$ (in C_{2v} geometry) from $H(^2S_g) + BH^+(^2\Sigma^+)$, the symmetry during formation is C_s. Resolution of both ^2S_g and $^2\Sigma^+$ into C_s symmetry gives A' and the symmetry of unsymmetric bent $[HBH]^+$ will be A'. However A' symmetry can be obtained from both the A_1 and B_2 representations in C_{2v} and it is not possible, without some further information, to decide which state of symmetric $[HBH]^+$ will be formed.

Determination of the molecular states arising from the combination of two like fragments is rather more straightforward, because the full symmetry of the final molecule may exist at large separations, even though the fragments may be of different symmetry. If the two fragments are in different electronic states one can consider either one as being excited and as the two fragments approach, one can construct two wavefunctions. One is symmetric and the other is antisymmetric with respect to the new element of symmetry introduced. Thus one gets twice as many states as are obtained when the fragments are in the same electronic state. Let us consider the approach of two NO_2 molecules and assume the formation of a stable dimer having a planar structure with D_{2h} symmetry. Let one NO_2 molecule be in the $\tilde{X}\,^2A_1$ state and the other in the first excited state $\tilde{A}\,^2B_1$. The direct product (in C_{2v} symmetry) is B_1 which can be obtained from both the representations B_{2g} and B_{3u} on resolution of D_{2h} into C_{2v}. Thus we will get four states of N_2O_4, namely $^3B_{2g}$, $^3B_{3u}$, $^1B_{2g}$ and $^1B_{3u}$. However if both NO_2 molecules are in the $\tilde{X}\,^2A_1$ state, we obtain only one singlet and one triplet state. Both the A_g and B_{1u} representations of D_{2h} transform to A_1 as the symmetry is lowered to C_{2v}. In order to decide which of these states are obtained in N_2O_4, it is necessary to consider the two fragments as united atoms (in ^2S states). The resulting diatomic molecule will have $^1\Sigma_g^+$ and $^3\Sigma_u^+$ states (from table 2.3). Resolving these into D_{2h} symmetry results in ^1A_g and $^3B_{1u}$ states, respectively.

3.3. Intersection of Triatomic Potential Surfaces

In section 2.7 we discussed the non-crossing rule for diatomic potential curves of the same spin and symmetry. In equations 2.46 to 2.48 we derived the energies for two functions ψ_1 and ψ_2 derived as linear combinations of two functions ϕ_1 and ϕ_2 which are mutually orthogonal and also orthogonal to all the electronic wavefunctions for the other states. The energies E are obtained by solving the secular determinant

$$\begin{vmatrix} H_{11} - E & H_{12} \\ H_{12} & H_{22} - E \end{vmatrix} = 0 \qquad (3.8)$$

where

$$H_{ij} = \int \phi_i H \phi_j \; d\tau,$$

to give

$$E = \tfrac{1}{2}(H_{11} + H_{22}) \pm \tfrac{1}{2}[(H_{11} - H_{22})^2 + 4H_{12}^2]^{\frac{1}{2}}. \qquad (3.9)$$

The two energies will be identical only if $H_{11} = H_{22}$ and $H_{12} = 0$. In a diatomic molecule there is only one geometrical parameter, the interatomic distance, and it is not possible to satisfy both conditions simultaneously. In the case of a polyatomic molecule there are more degrees of freedom and Teller (1937) showed that two surfaces may cross if it is possible to vary two parameters, say x and y. If the origin, i.e. the point at which the surfaces cross, is taken to be the point at which $H_{11} = H_{22}$ and $H_{12} = 0$, the secular determinant can be rewritten in the form

$$\begin{vmatrix} W + h_1 x - E & ly \\ ly & W + h_2 x - E \end{vmatrix} = 0 \qquad (3.10)$$

which is diagonal when $y = 0$. If we rewrite the diagonal elements in terms of $m = \frac{1}{2}(h_1 + h_2)$ and $k = \frac{1}{2}(h_1 - h_2)$ we obtain

$$\begin{vmatrix} W + (m + k)x - E & ly \\ ly & W + (m - k)x - E \end{vmatrix} = 0 \qquad (3.11)$$

which has the solutions

$$E = W + mx \pm (k^2 x^2 + l^2 y^2)^{\frac{1}{2}}. \qquad (3.12)$$

This is the equation of a double cone with the common vertex at the origin $x = 0$, $y = 0$. Thus at the point of intersection the potential energy surfaces, expressed as functions of x and y, form a double cone and intersect at a single point known as a conical intersection. This is illustrated in figure 3.10.

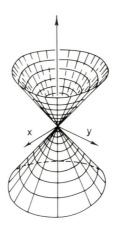

Figure 3.10. Schematic diagram of a conical intersection between two potential surfaces.

It has been shown (Herzberg and Longuet-Higgins 1963, Longuet-Higgins 1975) that the electronic wavefunction changes sign when one goes once round the conical intersection. In the general polyatomic case the potential function depends on more than two variables and thus the intersection between two surfaces will be a surface of dimension $m - 2$ where m is the number of variables in the potential function.

Molecules with C_{2v} symmetry provide examples of conical intersections. The ion NH_2^+ is of C_{2v} symmetry and has electronic states of $^3A_2$ and $^3B_1$ symmetry (with the plane of the molecule being the yz plane and the C_2 axis defined as the z-axis). Let us consider a geometry such that these two surfaces intersect. Any distortion from this geometry can be represented as a linear combination of the normal coordinates used in vibrational spectros-copy. There are three of these which can be represented by Q_1 (symmetric stretch), Q_2 (bending) and Q_3 (asymmetric stretch). Q_1 and Q_2 are totally symmetric, belonging to the irreducible representation A_1, and do not distort the symmetry from that of an isosceles triangle, whereas the asymmetric stretch is of B_2 symmetry and does distort the symmetry. If the geometry is varied in such a way that Q_3 is zero, the symmetry of the molecule does not change from C_{2v}, and there is a line of intersection between the $^3A_2$ and $^3B_1$ surfaces which will be a function of Q_1 and Q_2. However if Q_3 is non-zero, the symmetry is lowered to C_s and both surfaces are of $^3A''$ symmetry. Thus there will be a non-zero off-diagonal matrix element H_{12} which will be a function of Q_3. The secular determinant can be written in the form

$$\begin{vmatrix} H_{11}(Q_1, Q_2) & H_{12}(Q_3) \\ H_{12}(Q_3) & H_{22}(Q_1, Q_2) \end{vmatrix} = 0. \qquad (3.13)$$

If Q_1 is constrained to be zero, the energies of the two surfaces will be functions of Q_2 and Q_3 and will form a double cone with origin at $Q_2 = 0$, $Q_3 = 0$. Similarly for $Q_2 = 0$, there will be a conical intersection for the surfaces expressed in terms of Q_1 and Q_3.

Conical intersections can also arise for linear molecules when two degenerate states intersect. Consider the case of a linear triatomic molecule XYZ in which a Π state intersects a Δ state. If the molecule is bent, the symmetry is lowered from $C_{\infty v}$ to C_s and the degeneracy of each state is removed. The Π and Δ states each split into two states, one of A' symmetry and one of A'' symmetry. There will now be non-zero matrix elements between the two surfaces of A' symmetry and the intersection will be avoided. These two A' surfaces can intersect only when the geometry passes through the linear configuration. There is a conical intersection centred at that point (see figure 3.11). An example of this is the molecule HNO, for which the lowest singlet state, for linear geometries, is $^1\Delta$ (derived from a π^2 electronic configuration). There is a low-lying $^1\Pi$ state arising from either a $\pi\sigma$ or $\sigma\pi^3$ configuration. On consideration of the correlation rules we see that dissociation to HN + O or to H + NO yields the following fragments:

$$\text{HNO}(^1\Delta) \rightarrow \begin{cases} \text{HN}(A\,^3\Pi) + \text{O}(^3P) \\ \text{H}(^2S) + \text{NO}(B'\,^2\Delta) \end{cases} \qquad (3.14)$$

and

$$\text{HNO}(^1\Pi) \rightarrow \begin{cases} \text{HN}(X\,^3\Sigma^-) + \text{O}(^3P) \\ \text{H}(^2S) + \text{NO}(X\,^2\Pi) \end{cases}. \qquad (3.15)$$

Thus the $^1\Delta$ state yields excited states of HN or NO on dissociation, whereas the $^1\Pi$ state dissociates to ground-state fragments. For linear geometries the potential surfaces for the $^1\Delta$ and $^1\Pi$ states will cross when R_{NO} is increased, with R_{NH} kept fixed, or when R_{NH} is increased with R_{NO} held constant. On

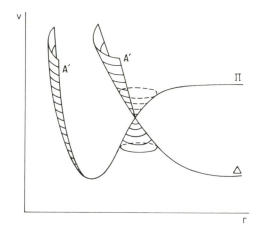

Figure 3.11. Conical intersection of the two A′ potential energy surfaces derived from a Π state and a Δ state of a linear molecule. (Adapted from Herzberg 1966.)

bending the molecule at the point of intersection, there will be separate conical intersections for the $^1A'$ and $^1A''$ states similar to that shown in figure 3.11.

Three-fold degenerate intersections can occur for linear configurations in which a Σ state intersects a Π or Δ state. On bending the Π or Δ state splits into A′ and A″ states whereas the Σ state is non-degenerate and becomes A′ or A″ depending on whether it is Σ^+ or Σ^-. The ground state of linear HOH (with bond lengths of the order of 1 Å) is $^1\Sigma_g^+$ and the first excited state is $^1\Pi$. However, on the basis of the correlation rules we see that the $^1\Sigma^+$ state dissociates to an excited state of $OH(A\,^2\Sigma^+)$, whereas the excited state $^1\Pi$ of HOH yields ground-state fragments

$$HOH(^1\Sigma^+) \rightarrow H(^2S) + OH(A\,^2\Sigma^+) \tag{3.16}$$

and

$$HOH(^1\Pi) \rightarrow H(^2S) + OH(X\,^2\Pi). \tag{3.17}$$

Thus, for linear configurations, as one of the OH bonds is stretched, the $^1\Sigma^+$ and $^1\Pi$ states must cross. The line of intersection between the two surfaces will be a function of the two OH distances R_{OH_1} and R_{OH_2}. Bending the molecule at a particular geometry for which the collinear surfaces intersect results in two $^1A'$ surfaces and one $^1A''$ surface. There is a conical intersection between the two $^1A'$ surfaces and the $^1A''$ surface will be tangential to the two cones at the point of intersection. A detailed discussion of intersections of potential surfaces for triatomic molecules has been given by Carter, Mills and Dixon (1984).

It might be argued that these are special cases for which there are particular geometries for which the symmetry is higher.than the general C_s symmetry for a triatomic species. The two surfaces of A′ and A″ symmetry correlate with different symmetries in the higher point group. The matrix element H_{12} is thus zero at this point and the two surfaces can intersect. However, Herzberg and Longuet-Higgins (1963) and Longuet-Higgins (1975) considered the case of a molecule made up from three unlike atoms, such as Li, Na and K in ^{2}S states. Such a molecule will have A′ symmetry for all geometries. Qualitative valence bond considerations indicate that there is a loop in configuration space such that the wavefunction of the lowest doublet state changes sign on traversing the loop. This implies the existence of a genuine intersection between two surfaces of the same symmetry.

Suggestions for Further Reading

T. Carrington (1974) *Acc. Chem. Research*, **7**, 20.
G. Herzberg (1966) *Molecular Spectra and Molecular Structure*, Vol. III. *Electronic Spectra and Electronic Structure of Polyatomic Molecules*, Van Nostrand, Princeton.
P. J. Kuntz (1976) in *Dynamics of Molecular Collisions*, ed. W. H. Miller, part B, p. 53, Plenum Press, New York.
J. N. Murrell (1978) in *Gas Kinetics and Energy Transfer*, Vol. 3, ed. P. G. Ashmore and R. J. Donovan, p. 200, The Chemical Society, London.

References

S. Carter, I. M. Mills and R. N. Dixon (1984) *J. Mol. Spect.*, **106**, 411.
R. J. Donovan and D. Husain (1970) *Chem. Rev.*, **70**, 489.
G. Herzberg (1966) *Molecular Spectra and Molecular Structure*, Vol. III. *Electronic Spectra and Electronic Structure of Polyatomic Molecules*, Van Nostrand, Princeton.
G. Herzberg and H. C. Longuet-Higgins (1963) *Disc. Farad. Soc.*, **35**, 77.
D. M. Hirst (1978) *Mol. Phys.*, **35**, 1559.
H. C. Longuet-Higgins (1975) *Proc. Roy. Soc. A*, **344**, 147.
B. H. Mahan (1971) *J. Chem. Phys.*, **55**, 1436.
J. N. Murrell (1980) *Israel J. Chem.*, **19**, 283.
E. Teller (1937) *J. Phys. Chem.*, **41**, 109.

the relationship between potential surfaces and vibrational spectra

4.1. Determination of the Potential Energy Surface from Spectroscopic Data

In chapter 2 we saw that the potential energy function for a diatomic molecule can be expanded in the form

$$V(R) = \frac{1}{2!}f_2(R-R_e)^2 + \frac{1}{3!}f_3(R-R_e)^3 + \frac{1}{4!}f_4(R-R_e)^4 + \ldots, \quad (4.1)$$

where f_2 is the harmonic force constant. The situation for a polyatomic molecule is much more complicated because, for a molecule containing N atoms, there are now $3N-6$ ($3N-5$ for a linear molecule) vibrational modes whereas a diatomic molecule has only a single mode. A general way of expressing the potential function for a polyatomic molecule, appropriate to configurations not too far displaced from equilibrium, is to write it as a Taylor series expansion in terms of displacement coordinates r_i from the equilibrium configuration

$$V = \frac{1}{2!}\sum_i\sum_j f_{ij}r_ir_j + \frac{1}{3!}\sum_i\sum_j\sum_k f_{ijk}r_ir_jr_k$$

$$+ \frac{1}{4!}\sum_i\sum_j\sum_k\sum_l f_{ijkl}r_ir_jr_kr_l + \ldots. \quad (4.2)$$

Here f_{ij} are the harmonic force constants and f_{ijk} and f_{ijkl} are the cubic and quartic force constants respectively. When the potential function is written in this way, the force constants are equal to the derivatives of the potential evaluated at the equilibrium configuration, e.g.

$$f_{ijk} = \left(\frac{\partial^3 V}{\partial r_i \partial r_j \partial r_k}\right)_e \quad (4.3)$$

The potential function does not contain linear terms because the potential energy is a minimum at the equilibrium configuration and thus the first derivatives $(\partial V/\partial r_i)_e$ are zero. The Taylor series expansion of equation 4.2 converges fairly rapidly for displacements corresponding to vibrational excitation of only one or two quanta, and in general it is not possible to determine terms in the potential function beyond the quartic term. It should be noted that in the literature some authors define force constants which may differ from the above convention by a numerical factor (the factorial term) or in the way in which the summations are carried out. Thus the reader must be careful to establish exactly which convention is being used. Also there is some confusion with regard to units, particularly as the SI scheme has not been universally adopted by spectroscopists. Bond-stretching coordinates are usually written in terms of ångströms ($1\,\text{Å} = 10^{-10}\,\text{m}$) and stretching force constants are expressed as follows: quadratic—$\text{mdyn}\,\text{Å}^{-1}$; cubic—$\text{mdyn}\,\text{Å}^{-2}$; quartic—$\text{mdyn}\,\text{Å}^{-3}$ ($1\,\text{dyne} = 10^{-5}\,\text{N}$). Alternatively, quadratic force constants can be written in terms of $\text{aJ}\,\text{Å}^{-2}$ because $1\,\text{mdyn}\,\text{Å}^{-1} = 1\,\text{aJ}\,\text{Å}^{-2}$. Angle-bending coordinates are expressed either in radians or in ångströms by the scaling of one of the bonds on either side of the angle by the equilibrium bond length. It is, however, becoming increasingly common for force constants to be expressed in terms of $\text{N}\,\text{m}^{-1}$.

4.1.1. The Harmonic Force Field

We consider first the case of the harmonic potential function where the cubic and quartic terms in equation 4.2 are neglected. Thus we seek a potential energy function of the form

$$V = \frac{1}{2!} \sum_i \sum_j f_{ij} r_i r_j. \qquad (4.4)$$

For some molecules it is not at present possible to go beyond the harmonic potential function. The cubic and quartic anharmonic force constants can only be determined from a detailed analysis of vibration–rotation interactions, and for many molecules the data are not available at present, so that only the harmonic potential function can be determined. For other molecules where a more complete treatment is possible, the calculation starts with the determination of the harmonic potential function. It is necessary first of all to define an appropriate coordinate system. It is usual to start with the assumption that the vibrational motion is separable from the translational and rotational motion of the molecule by introducing a set of molecule fixed axes X, Y, Z located at the centre of mass of the molecule and considering only the motion of the atoms relative to these axes. This is usually done by use of the Eckart conditions to obtain, for a non-linear molecule, $3N-6$ internal coordinates D_i such that the linear momentum is zero in each of the mutually orthogonal directions X, Y, Z and also such that the vibrational angular momentum is minimized with respect to this coordinate system. For a fuller description the reader is referred to Wilson et al. (1955). Critical appraisals of the procedure have been given by Sutcliffe (1980, 1982).

If the mass of each atom is m_i and its equilibrium position is given by X_i^e, Y_i^e, Z_i^e, then the former condition leads to

$$\sum_i m_i x_i = \sum_i m_i y_i = \sum_i m_i z_i = 0, \qquad (4.5)$$

where x_i, y_i, z_i are the displacements of atom i relative to the axes X, Y, Z. Applying the second condition of minimum angular momentum relative to the molecule fixed axes yields the conditions

$$\sum_i m_i(Y_i^e z_i - Z_i^e y_i) = 0,$$

$$\sum_i m_i(Z_i^e x_i - X_i^e z_i) = 0, \qquad (4.6)$$

$$\sum_i m_i(X_i^e y_i - Y_i^e x_i) = 0.$$

Equations 4.5 and 4.6 constitute the Eckart conditions and they are sufficient to locate the molecule fixed axes X, Y, Z on the moving (vibrating–rotating) molecule.

If we consider a linear triatomic molecule ABC, the chemically intuitive internal coordinates are the AB and BC stretches and the ABC angle bend, which is twofold degenerate. Choosing internal coordinates in terms of geometrically defined bond-stretching and angle-bending coordinates leads to harmonic force constants which are independent of isotopic substitution. In general one has to be careful to ensure that the set of internal coordinates does not contain any redundant coordinates. Let the internal coordinates for the ABC molecule be denoted by d_1, d_2, d_3 where

$$d_1 = \Delta R_{AB}, \qquad d_2 = \Delta R_{BC}, \qquad d_3 = (R_{AB}R_{BC})^{\frac{1}{2}}\Delta\theta. \qquad (4.7)$$

In terms of Cartesian displacement coordinates relative to the molecule fixed axes X, Y, Z (with Z along the internuclear axis) the stretching coordinates are

$$d_1 = z_B - z_A, \qquad d_2 = z_C - z_B, \qquad (4.8)$$

where z_A, z_B, z_C are the displacements, along the Z axis, of nuclei A, B, C. If we impose the condition

$$m_A z_A + m_B z_B + m_C z_C = 0 \qquad (4.9)$$

we obtain the relationships

$$z_A = -\frac{m_B + m_C}{M}d_1 - \frac{m_C}{M}d_2,$$

$$z_B = \frac{m_A}{M}d_1 - \frac{m_C}{M}d_2, \qquad (4.10)$$

Potential Energy Surfaces

$$z_C = \frac{m_A}{M} d_1 + \frac{m_A + m_B}{M} d_2,$$

where $M = m_A + m_B + m_C$. The centre of mass of the molecule lies along the molecular (Z) axis and equation 4.5 is clearly satisfied. Let us now consider bending in the XZ plane in which the nuclei undergo infinitesimal displacements x_A, x_B, x_C as indicated in figure 4.1.

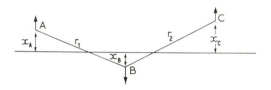

Figure 4.1. Displacement coordinates for bending in XZ plane.

The change in angle, for infinitesimal displacements, can be written

$$\Delta\theta = \frac{x_B - x_A}{r_1} + \frac{x_B - x_C}{r_2} \tag{4.11}$$

and thus

$$d_{3x} = \left(\frac{r_2}{r_1}\right)^{\frac{1}{2}} (x_B - x_A) + \left(\frac{r_1}{r_2}\right)^{\frac{1}{2}} (x_B - x_C). \tag{4.12}$$

Imposing the conditions

$$m_A x_A + m_B x_B + m_C x_C = 0 \tag{4.13}$$

and (from equation 4.6)

$$m_A Z_A^e x_A + m_B Z_B^e x_B + m_C Z_C^e x_C = 0, \tag{4.14}$$

where Z_A^e, Z_B^e, Z_C^e are the equilibrium Z-coordinates of nuclei A, B, C relative to the centre of mass, leads to the following expressions for x_A, x_B, x_C in terms of d_{3x}:

$$x_A = -\frac{m_B m_C r_2^{\frac{1}{2}}}{N r_1^{\frac{1}{2}}} d_{3x},$$

$$x_B = \frac{m_A m_C (r_1 + r_2)}{N r_1^{\frac{1}{2}} r_2^{\frac{1}{2}}} d_{3x}, \tag{4.15}$$

$$x_C = -\frac{m_A m_B r_1^{\frac{1}{2}}}{N r_2^{\frac{1}{2}}} d_{3x},$$

where

$$N = \frac{m_A m_B r_1^2 + m_A m_C (r_1 + r_2)^2 + m_B m_C r_2^2}{r_1 r_2}.$$

Corresponding expressions can be obtained for y_A, y_B, y_C in terms of d_{3y} for bending in the YZ plane. The transformation from displacement coordinates x_A, y_A, z_A etc. to internal coordinates d_i can be written in matrix notation, thus:

$$\begin{pmatrix} D \\ 0 \end{pmatrix} = \begin{pmatrix} B \\ \beta \end{pmatrix} (x), \tag{4.16}$$

where x is a column vector containing the Cartesian displacement coordinates and D represents the $3N-6$ internal coordinates. The portion β of the transformation matrix represents the six Eckart conditions, i.e., $[\beta][x] = 0$ comprises the six equations 4.5 and 4.6.

The vibrational kinetic energy T of the molecule is given by

$$2T = \sum_{i=A,B,C} m_i(\dot{x}_i^2 + \dot{y}_i^2 + \dot{z}_i^2), \tag{4.17}$$

where $\dot{x}_i$ represents the time derivative of x_i etc. Substitution of equations 4.10 and 4.15 for the Cartesian displacements in terms of internal coordinates gives

$$2T = \frac{m_A(m_B + m_C)}{M} \dot{d}_1^2 + \frac{m_C(m_A + m_B)}{M} \dot{d}_2^2$$
$$+ \frac{2m_A m_C}{M} \dot{d}_1 \dot{d}_2 + \frac{m_A m_B m_C}{N} (\dot{d}_{3x}^2 + \dot{d}_{3y}^2), \tag{4.18}$$

where $M = m_A + m_B + m_C$. The kinetic energy is thus of the general form

$$T = \tfrac{1}{2} \sum_i \sum_j \mu_{ij} \dot{d}_i \dot{d}_j, \tag{4.19}$$

where the μ_{ij} are the coefficients of the $\dot{d}_i \dot{d}_j$ in equation 4.18. The potential energy for this molecule will have the same general form, namely

$$2V = f_{11} d_1^2 + f_{22} d_2^2 + f_{12} d_1 d_2 + f_{21} d_2 d_1 + f_{33}(d_{3x}^2 + d_{3y}^2). \tag{4.20}$$

The terms involving $d_1 d_3$ and $d_2 d_3$ can be shown to be zero by the requirement that the potential energy is invariant with respect to symmetry operations of the molecule. Similarly, $f_{12} = f_{21}$. In the general case

$$V = \tfrac{1}{2} \sum_i \sum_j f_{ij} d_i d_j. \tag{4.21}$$

Corresponding expressions can be derived for other molecules. For symmetric molecules (e.g. a bent XYX molecule) it is easier to obtain symmetry coordinates S_i which are combinations of the displacement coordinates which have the symmetry of the molecule. If S_i and S_j belong to different symmetry species, then the corresponding force constant f_{ij} is zero. Thus if for each symmetry species there are n_k vibrations, each degenerate vibration being counted once only, then the total number of independent force constants is given by

$$\sum_k \tfrac{1}{2} n_k (n_k + 1),$$

where the summation runs over all symmetry species.

We now have to relate the force constants f_{ij} in the potential function V (equation 4.21) to the observed vibration wavenumbers ω_i of the molecule. In order to do this it will be necessary to transform the internal coordinates to normal coordinates. This procedure is most clearly introduced through a classical discussion of the vibrational motion. The importance of normal coordinates is that in this coordinate system the vibrational Schrödinger equation, which is of dimension $3N-6$, is separable into a set of one-dimensional equations. However before proceeding to derive the normal coordinates we note at this point that for a triatomic molecule there are, in general, four force constants f_{11}, f_{12}, f_{22} and f_{33} but only three vibrational wavenumbers to be observed.

In classical terms, the vibrational motion of a molecule in the harmonic approximation is regarded as the superposition of motion in $3N-6$ normal modes in which each atom executes simple harmonic motion. In each normal mode all the particles move with the same frequency ν and, in general, in phase but with different amplitudes. This can be generalized to the quantum mechanical treatment in which the motion is regarded as the superposition of a set of quantum mechanical oscillators. Thus we can write for each internal coordinate

$$d_i = A_i \cos(\lambda^{\frac{1}{2}} + \epsilon), \tag{4.22}$$

where A_i is the amplitude of the motion, ϵ is the phase factor and

$$\lambda = 4\pi^2 \nu^2 = 4\pi^2 c^2 \omega^2 \tag{4.23}$$

in which ν is the vibration frequency (in hertz) and ω is the vibration wavenumber (usually expressed in units of cm^{-1}). Using this assumption we must next obtain the classical equations of motion by using Lagrange's equation which is of the form

$$\frac{d}{dt}\left(\frac{\partial T}{\partial \dot{x}_i}\right) + \frac{\partial V}{\partial x_i} = 0, \quad \text{for } i = 1, \ldots, 3N, \tag{4.24}$$

for a system where the kinetic energy T is of the form

$$2T = \sum_i \sum_j \mu_{ij} \dot{x}_i \dot{x}_j \tag{4.25}$$

and the potential energy V is a function of the coordinates x_i. It is not appropriate to discuss the Lagrangian formulation of classical mechanics here but the reader will recognize that equation 4.24 is equivalent to the Newtonian relation between force and acceleration. The derivative of the potential, $\partial V / \partial x_i$, is equal to and opposite to the force acting on the particle in the direction of x_i, while the time derivative is the derivative of the momentum in the direction of x_i. Thus we obtain three equations of motion by considering d_1, d_2, and d_3 in turn

$$\mu_{11}\ddot{d}_1 + \mu_{12}\ddot{d}_2 + f_{11}d_1 + f_{12}d_2 = 0$$
$$\mu_{12}\ddot{d}_1 + \mu_{22}\ddot{d}_2 + f_{12}d_1 + f_{22}d_2 = 0 \tag{4.26}$$
$$\mu_{33}\ddot{d}_3 + f_{33}d_3 = 0.$$

The bending mode is two-fold degenerate so it is only necessary to consider d_{3x} or d_{3y}. Substituting for d_1, d_2 and d_3 from equation 4.22 leads to the following set of homogeneous equations for the amplitudes A_i

$$(f_{11} - \lambda\mu_{11})A_1 + (f_{12} - \lambda\mu_{12})A_2 = 0$$
$$(f_{12} - \lambda\mu_{12})A_1 + (f_{22} - \lambda\mu_{22})A_2 = 0 \tag{4.27}$$
$$(f_{33} - \lambda\mu_{33})A_3 = 0.$$

These equations will only have a non-trivial solution (i.e., other than $A_1 = A_2 = A_3 = 0$) if the determinant of the coefficients of the amplitudes A_i is equal to zero, i.e.,

$$\begin{vmatrix} f_{11} - \lambda\mu_{11} & f_{12} - \lambda\mu_{12} & 0 \\ f_{12} - \lambda\mu_{12} & f_{22} - \lambda\mu_{22} & 0 \\ 0 & 0 & f_{33} - \lambda\mu_{33} \end{vmatrix} = 0. \tag{4.28}$$

In the determinant, the terms μ_{ij} are functions of the masses of the atoms and are thus known. The terms f_{ij} are the force constants for the potential function expressed in terms of internal coordinates and λ is related to the vibration wavenumbers ω_i. It is for those values of $\lambda = 4\pi^2c^2\omega_i^2$ that we require the determinant to be equal to zero and it is by using the experimental vibration wavenumbers that we hope to be able to calculate the force constants f_{ij}. This is a straightforward procedure for the bending mode, which is conventionally labelled ν_2, because the condition is satisfied by

$$\lambda_2 = \frac{f_{33}}{\mu_{33}} \tag{4.29}$$

and the force constant can be obtained directly. However, for the stretching modes there are only two experimental vibration wavenumbers available for the determination of the force constants f_{11}, f_{12}, f_{22} for an ABC molecule.

We defer discussion of this difficulty and consider the reverse calculation in which we take a set of force constants f_{ij} and calculate the corresponding values of λ, and hence the vibration wavenumbers ω from equation 4.28. Let us assume that there is a set of values f_{11}, f_{12}, f_{22} for which the determinant

$$\begin{vmatrix} f_{11}-\lambda\mu_{11} & f_{12}-\lambda\mu_{12} \\ f_{12}-\lambda\mu_{12} & f_{22}-\lambda\mu_{22} \end{vmatrix} \qquad (4.30)$$

is zero for the values of λ derived from the two experimental stretching vibration wavenumbers. For example, a harmonic force field calculation for HCN (with the labelling scheme H = A, C = B, N = C) yielded the force constants $f_{11} = 6.251\,\text{mdyn Å}^{-1}$, $f_{12} = -0.200\,\text{mdyn Å}^{-1}$ and $f_{22} = 18.730\,\text{mdyn Å}^{-1}$. To convert to SI units (N m^{-1}) it is necessary to multiply by 10^2. Using the atomic weights H = 1.0078 u, C = 12.0 u and N = 14.0031 u [the unified atomic mass unit, 1 u ($\frac{1}{12}$ of the mass of the C atom) = 1.6605×10^{-27} kg] we obtain the determinant

$$\begin{vmatrix} 625.1-1.6110\times10^{-27}\lambda & -20.0-0.8676\times10^{-27}\lambda \\ -20.0-0.8676\times10^{-27}\lambda & 1870.3-11.1976\times10^{-27}\lambda \end{vmatrix} = 0. \quad (4.31)$$

It is convenient to express λ in terms of the harmonic vibration wavenumber ω using the relationship $\lambda\,/\,\text{kg}^{-1}\,\text{J m}^{-2} = 3.5483\times10^{22}(\omega\,/\,\text{cm}^{-1})^2$ and rewrite the determinant as

$$\begin{vmatrix} 625.1-57.1631x & -20.0-30.7850x \\ -20.0-30.7850x & 1870.3-397.3313x \end{vmatrix} = 0, \qquad (4.32)$$

where $\sqrt{x} = \omega\times10^{-3}\,/\,\text{cm}^{-1}$. The determinant can be expanded as a quadratic equation which can be solved to give the harmonic stretching wavenumbers $3442\,\text{cm}^{-1}$ and $2129\,\text{cm}^{-1}$.

We can now return to equation 4.27 to obtain the amplitudes A_1, A_2 and A_3 for each of the three normal modes. We designate these as A_{ik}, where k labels the mode in question. Clearly for the bending mode (conventionally labelled ν_2), $A_{12} = A_{22} = 0$ and A_{32} can be set to unity because it is conventional to normalize the amplitudes A_{ik} such that

$$\sum_i A_{ik}^2 = 1. \qquad (4.33)$$

For mode ν_1 with $\omega = 2129\,\text{cm}^{-1}$ we obtain the equations

$$366.0254A_{11} - 159.5238A_{21} = 0,$$

$$-159.5238A_{11} + 69.5151A_{21} = 0. \qquad (4.34)$$

Thus $A_{21} = 2.2945A_{11}$ and we normalize by dividing through by

$$\left(\sum_i A_{ik}^2 \right)^{\frac{1}{2}}$$

to yield $A_{11} = 0.3995$, and $A_{21} = 0.9167$. For mode ν_3 (with $\omega = 3442\,\text{cm}^{-1}$) the normalized amplitudes are $A_{13} = 0.9906$, and $A_{23} = -0.1365$. The dominant contribution to the mode ν_1 with vibration wavenumber $2129\,\text{cm}^{-1}$ is d_2 and this mode is thus primarily a CN stretching mode. Similarly ν_3 corresponds to CH stretching.

In addition to giving the relative amplitudes for the bond stretches d_1 and d_2 in normal modes ν_1 and ν_3, the coefficients A_{ik} enable us to transform the internal coordinates d_i to the normal coordinates Q_k for the molecule. The transformation is given by

$$Q_k = \sum_i A_{ik} d_i. \tag{4.35}$$

The reverse transformation is

$$d_i = \sum_k A_{ik} Q_k, \tag{4.36}$$

which is conventionally written in matrix notation as

$$\boldsymbol{D} = \boldsymbol{LQ}, \tag{4.37}$$

where $\boldsymbol{L}$ is a matrix whose elements are the normalized amplitudes A_{ik}. The importance of this transformation becomes apparent when the kinetic energy and potential energy are expressed in terms of the normal coordinates. The normal coordinates are defined by requiring diagonal expressions for T and V, with all effective masses equal to 1, in normal coordinate space

$$2T = \sum_k \dot{Q}_k^2 \tag{4.38}$$

and

$$2V = \sum_k \lambda_k Q_k^2. \tag{4.39}$$

Thus the energy expressions are sums of terms involving squares of $\dot{Q}_k$ or Q_k and there are no cross terms coupling motions in different modes. The classical equations of motion for the molecule can be written

$$\sum_k (\ddot{Q}_k - \lambda_k Q_k) = 0. \tag{4.40}$$

Transformation to quantum mechanics yields a Schrödinger equation of dimension $3N-6$ which can be separated into $3N-6$ one-dimensional equations.

The example discussed above is relatively simple. For larger molecules more sophisticated methods are needed for the determination of the terms μ_{ij} in equations 4.19, 4.26 and 4.27. Equation 4.28 can be rewritten in matrix notation as

$$F - G^{-1}\lambda = 0, \tag{4.41}$$

where F is a matrix of the force constants and G^{-1} is the matrix of the mass dependent terms μ_{ij}. This equation can be rewritten as

$$GF - \lambda E = 0, \tag{4.42}$$

where E is the unit matrix. If the molecule has symmetry, it is advantageous to transform the internal coordinates D to symmetry coordinates S which transform as the various irreducible representations of the molecular point group. The G matrix is then formulated in terms of the symmetry coordinates. This has the considerable advantage that the matrix $GF - \lambda E$ assumes block diagonal form with subsequent simplification of the eigenvalue calculation. Also, as we saw above in the linear ABC case, it is only necessary to consider one component of a set of degenerate modes. Elegant methods for the calculation of the G matrix have been derived by Wilson but a discussion of these methods is beyond the scope of this book. For further information the reader is referred to the books by Wilson *et al.* (1955) and by Woodward (1972).

We return now to the problem that in general there are insufficient experimental vibrational wavenumbers ω_k for the determination of the force constants f_{ij} in the harmonic force field

$$V = \tfrac{1}{2} \sum_i \sum_j f_{ij} d_i d_j. \tag{4.43}$$

There are two ways of approaching this problem. The first approach is to make some simplifying assumption and thus reduce the number of parameters to be determined. The second, and more satisfying, method is to look for experimental parameters, other than the vibration wavenumbers, that depend on the potential energy function and to use such additional information in the determination of the full harmonic potential function or force field.

A potential function of the form of equation 4.43 in which the displacement coordinates d_i are in terms of bond stretching and angle bending is known as a general valence force field. One of the simplest assumptions to make is that the off-diagonal force constants f_{ij} ($i \neq j$) are so small that they can be set to zero. This approximation is known as the simple valence force field. If the number of force constants to be determined is less than the number of experimental vibration frequencies, it is then possible to get some idea of the adequacy of the force field from the accuracy with which it reproduces the vibration frequencies not used in its determination.

Another type of function which has been widely used is the Urey–Bradley potential function which includes some interaction terms between non-bonded atoms in addition to the usual bond stretch and angle bend interactions. The introduction of the non-bonded interaction requires initially the presence of linear terms in the potential function which is of the form

$$V = \sum_i k_i r_i \Delta r_i + \sum_i l_i r_i \Delta \theta_i + \sum_i m_i \rho_i \Delta \rho_i$$
$$+ \tfrac{1}{2} \sum_i K_i (\Delta r_i)^2 + \tfrac{1}{2} \sum_i L_i (r_i \Delta \theta_i)^2 + \tfrac{1}{2} \sum_i M_i (\Delta \rho_i)^2, \quad (4.44)$$

where Δr_i, $\Delta \theta_i$ and $\Delta \rho_i$ represent changes in bond length r_i, bond angle θ_i and in the distance ρ_i between the two non-bonded atoms. It is then necessary to express $\Delta \rho_i$ in terms of Δr_i and $\Delta \theta_i$ and to eliminate the linear terms by imposing the condition that the potential V is a minimum at the equilibrium configuration. The final result of a considerable amount of algebra is a potential function containing quadratic terms but generally involving fewer force constants than the generalized valence force field. The Urey–Bradley force field is treated in more detail by Wilson *et al.* (1955), Woodward (1972) and Shimanouchi (1970).

Nevertheless, simplified force fields cannot be regarded as being completely satisfactory, however good the agreement they give with observed vibration frequencies. One should aim to determine all of the parameters in the harmonic force field, particularly as it is only the first approximation to the potential function for the molecule. We thus have to consider what additional pieces of information can be used in the calculation of the potential. In rotational spectroscopy, isotopic substitution is used in order to determine the geometries of polyatomic molecules. Since the potential function is independent of the nature of the isotopes of the atoms, we may ask if any further information can be obtained using isotopic substitution in vibrational spectroscopy. Isotopic substitution does give additional information, but not all of the frequencies in the substituted molecule provide new data because there are relationships, such as the Teller–Redlich product rules, between some of the vibrational frequencies in two isotopic variants. Also, the form of the normal coordinates usually does not change very much on isotopic substitution. The most favourable case is deuterium substitution where there is a large percentage change in mass. Nevertheless, some information is obtained which, in favourable cases, can be used in the determination of the force field. If we consider HCN and DCN, equation 4.30 gives two relationships between λ_1 and λ_2, for each molecule, and f_{11}, f_{12} and f_{22}. Thus by eliminating f_{22} a relationship between f_{11} and f_{12} can be obtained for each molecule. These functions relate pairs of numbers f_{11}, f_{12} which are consistent with the two values λ_1, λ_2 derived from the observed vibration wavenumbers. The functions for the two isotopic forms are found to intersect at two points, giving two sets of numbers $(f_{11}, f_{12})_1$ and $(f_{11}, f_{12})_2$ which satisfy the relationships for both molecules. Usually one of these sets gives force constants which are inconsistent with the nature of the molecule so the remaining pair gives the required values of f_{11} and f_{12}.

When other spectroscopic parameters are incorporated into the determination of the force field, the calculation becomes much more complicated because it involves additional equations in terms of partial derivatives of the parameter in question with respect to the force constants. Three types of spectroscopic information are routinely used to supplement the observed vibration wavenumbers. Although the vibration wavenumbers from isotopic variants of the molecule are of limited use in the direct determination of force constants, isotope shifts in the vibration wavenumbers can be related to changes in the force constant. Another important source of information is the Coriolis interaction between vibration and rotation in a polyatomic molecule. Consider a linear ABC molecule which is vibrating in the asymmetric stretching mode. On rotation about an axis passing through the centre of mass, additional forces, perpendicular to the molecular axis, come into play which correspond to the excitation of the bending mode. From a detailed analysis of the spectrum it is possible to obtain Coriolis coupling constants, which also depend on the force constants. The third effect that can be used to give further information about the force constants is the centrifugal distortion which gives rise to terms depending on $J^2(J+1)^2$ in the expression for the rotational energy. The energy-level expression for vibration–rotation levels contains terms analogous to the term $D_v J^2(J+1)^2$ in equation 2.23 for a diatomic molecule. The centrifugal distortion constant also depends on the force constants. Other information that can occasionally be used in the force field calculation includes the mean-square amplitudes of vibration obtained from electron diffraction studies, the inertial defect and intensity data.

The calculation of the force field using all the available data is a very difficult procedure involving an iterative method. Since the harmonic force field is really only an approximation to the actual force field, it should be regarded as the starting point for the calculation of the anharmonic force field. In the determination of the harmonic contribution to the total force field, the observed vibration wavenumbers are not the most appropriate starting point. In a diatomic molecule the fundamental vibration wavenumber is given by $\omega_e - 2x$, whereas the harmonic vibration wavenumber is ω_e. Thus for a polyatomic molecule it is necessary to analyse the overtone bands of the spectrum and from the anharmonicity constants deduce a set of harmonic vibration wavenumbers. If this is not possible an approximate correction can be applied to the fundamental vibration wavenumbers.

The iterative procedure takes a set of approximate force constants and calculates from them the values of the vibration wavenumbers and of the other experimental parameters used in the calculation. Derivatives of the vibration wavenumbers with respect to the force constants ($\partial \omega_k / \partial f_{ij}$) are obtained and used in a set of approximate linear equations relating changes in the force constants with changes in the vibration wavenumbers

$$\delta \omega_k = \sum_i \sum_j \left(\frac{\partial \omega_k}{\partial f_{ij}} \right) \delta f_{ij}. \qquad (4.45)$$

From the deviations between the observed data and the calculated values, changes in the force constants δf_{ij} are estimated and a refined set of force constants determined. The procedure is then repeated until the best weighted least-squares fit is obtained between the observed data and the calculated constants. The procedure seeks to minimize the sum

$$\sum w_i (a_{i(\text{obs})} - a_{i(\text{calc})})^2, \qquad (4.46)$$

where $a_{i(\text{obs})}$ and $a_{i(\text{calc})}$ are the observed and calculated values and w_i is the weight. Convergence is not easy to achieve and great care must be taken to ensure that the calculation does not diverge. For further details of the procedure the reader is referred to the book by Gans (1971) and articles by Duncan (1975) and Shimanouchi (1970).

4.1.2. Determination of the Anharmonic Force Field

The derivation of the anharmonic force field from spectroscopic data is a much more complicated procedure than the harmonic force field calculation outlined above. Here we only give an outline of the procedure and the reader is referred to Hoy *et al.* (1972) and Mills (1974) for more complete treatments.

The first problem is that the internal coordinates D (or S) used in the harmonic case are not adequate for the anharmonic force field calculation. Since we require the force field to be independent of the isotopic nature of the molecule it is important that the coordinate system should be defined geometrically and not contain any mass dependence. It is only in this way that we will be able to obtain cubic and quartic force constants which are isotopically invariant. It turns out that because the Eckart conditions involve the masses of the atoms, the coordinates D do have a mass dependence that becomes apparent beyond the harmonic approximation. It is thus necessary to transform to a set of internal coordinates (which, following Hoy *et al.* (1972), will be denoted Я) which are true curvilinear bond-stretching and angle-bending coordinates. They are identical with D_i (or S_i) for infinitesimal distortions but for larger displacements the bond lengths, in the system $\mathfrak{R}_i$, remain unchanged, whereas with D_i (or S_i) there is some bond stretching. In $\mathfrak{R}_i$ the atoms will move along curved paths whereas the coordinates D_i (or S_i) are rectilinear. In addition to resulting in mass-independent cubic and quartic force constants, the coordinates $\mathfrak{R}_i$ also lead to a simpler expression for the force field. We have previously derived the normal coordinates Q_i which are related to the internal coordinates D_i by the transformation

$$D_i = \sum_r L_i^r Q_r \tag{4.47}$$

or

$$\mathbf{D} = \mathbf{L}\mathbf{Q} \tag{4.48}$$

and the potential energy is of the form

$$V = \tfrac{1}{2} \sum_i \lambda_i Q_i^2. \tag{4.49}$$

The transformation to true curvilinear coordinates can be written

$$\mathfrak{R}_i = \sum_r L_i^r Q_r + \tfrac{1}{2} \sum_r \sum_s L_i^{rs} Q_r Q_s + \tfrac{1}{6} \sum_r \sum_s \sum_t L_i^{rst} Q_r Q_s Q_t, \tag{4.50}$$

where L_i^r are defined as before. The coefficients L_i^{rs} and L_i^{rst} are known as the second and third derivative elements of the $\mathbf{L}$ tensor and closed analytical expressions have been given for them by Hoy *et al.* (1972). The potential energy function is then written as

$$V = \tfrac{1}{2} \sum_i \sum_j f_{ij} \mathfrak{R}_i \mathfrak{R}_j + \tfrac{1}{6} \sum_i \sum_j \sum_k f_{ijk} \mathfrak{R}_i \mathfrak{R}_j \mathfrak{R}_k$$
$$+ \tfrac{1}{24} \sum_i \sum_j \sum_k \sum_l f_{ijkl} \mathfrak{R}_i \mathfrak{R}_j \mathfrak{R}_k \mathfrak{R}_l + \dots . \tag{4.51}$$

If we introduce dimensionless normal coordinates q_i by the relationship

$$q_i = \gamma_i^{1/2} Q_i, \tag{4.52}$$

where

$$\gamma_i = \frac{2\pi \lambda_i^{1/2}}{h}, \tag{4.53}$$

we obtain the following expression for the potential energy

$$\frac{V}{hc} = \tfrac{1}{2} \sum_i \omega_r q_i^2 + \tfrac{1}{6} \sum_i \sum_j \sum_k \phi_{ijk} q_i q_j q_k$$
$$+ \tfrac{1}{24} \sum_i \sum_j \sum_k \sum_l \phi_{ijkl} q_i q_j q_k q_l + \dots . \tag{4.54}$$

If the molecule has symmetry, the problem is formulated in terms of symmetrized curvilinear coordinates.

The calculation of anharmonic force fields is a much bigger task than the harmonic calculation because there are many more force constants to be determined. In the case of HCN there are six cubic and nine quartic force constants. The number of force constants escalates as the complexity of the molecule increases. In order to determine the force field it is necessary to use a larger number of spectroscopic parameters. Those used include the harmonic and anharmonic vibrational constants, rotational constants, Coriolis coupling constants, l-doubling constants, vibration–rotation interaction constants, centrifugal distortion constants and sextic distortion constants. In order to use these data it is necessary to obtain the derivatives of the constants with respect to the anharmonic force constants. The procedure employed is that of successive refinement (as in the harmonic case) in a large least-squares calculation. In view of the complexity of the calculation and the convergence difficulties inherent in it, the refinement is usually carried out in several stages as follows. From the rotational constants, the equilibrium structure (or some approximation to it) is first calculated. With the equilibrium structure constrained, the harmonic frequencies, Coriolis constants and centrifugal distortion constants are then used in the determination of the harmonic force field. The cubic force constants are then determined using the vibration–rotation interaction constants with the equilibrium structure and harmonic force constant constrained. Finally, the quartic force constants are determined using the anharmonic constants in a calculation in which the equilibrium structure and quadratic and cubic force constants are constrained.

4.2. Calculation of the Vibration–Rotation Energy Levels of a Polyatomic Molecule from the Potential Energy Surface

4.2.1. The Hamiltonian Operator

If one wishes to calculate the vibration–rotation levels of a polyatomic molecule, it is first necessary to define the Hamiltonian operator. One of the most widely used forms for non-linear molecules is that of Watson (1968), which is written

$$H = \tfrac{1}{2} \sum_{\alpha} \sum_{\beta} (\Pi_\alpha - \pi_\alpha)\mu_{\alpha\beta}(\Pi_\beta - \pi_\beta)$$

$$-\sum_{\alpha} \frac{h^2}{32\pi^2}\mu_{\alpha\alpha} + \sum_{k=1}^{3N-6} \left(-\frac{h^2}{8\pi^2}\frac{\partial^2}{\partial Q_k^2}\right) + V(Q). \qquad (4.55)$$

In this expression Q_k are the normal coordinates and $V(Q)$ is the potential energy function written in terms of these normal coordinates. The first term consists of the rotational kinetic energy and contributions from the interaction between vibration and rotation. The second term is a small mass-dependent correction to the effective potential and the third term represents the vibrational kinetic energy. The operator for the total angular momentum with respect to the molecule fixed axis α is given by Π_α and π_α is the vibrational angular momentum, along axis α, defined by

$$\pi_\alpha = \sum_k \sum_l \zeta_{kl}^\alpha Q_k \left(-\frac{ih}{2\pi} \frac{\partial}{\partial Q_l} \right) \tag{4.56}$$

in terms of the Coriolis coupling constants ζ_{kl}^α which are in turn related to the transformation coefficients $l_{\alpha i, k}$ used in the transformation from mass weighted Cartesian displacement coordinates to the normal coordinates

$$\zeta_{kl}^\alpha = \epsilon_{\alpha\beta\gamma} \sum_i l_{\beta i, k} l_{\gamma i, l} \tag{4.57}$$

and

$$Q_k = \sum_{\alpha = X, Y, Z} \sum_i l_{\alpha i, k} m_i^{1/2} (r_{\alpha i} - r_{\alpha i}^0), \tag{4.58}$$

where $r_{\alpha i} - r_{\alpha i}^0$ represents the displacement of nucleus i along axis α. The quantities $\mu_{\alpha\beta}$ are defined in terms of the elements of the inertia tensor $I_{\alpha\beta}$ by the relationship

$$\mu_{\alpha\beta} = (I'^{-1})_{\alpha\beta}, \tag{4.59}$$

where

$$I'_{\alpha\beta} = I_{\alpha\beta} - \sum_k \sum_l \sum_m \zeta_{km}^\alpha \zeta_{lm}^\beta Q_k Q_l. \tag{4.60}$$

This form of the Hamiltonian cannot be used for linear molecules because one of the components of μ diverges as the molecule becomes linear. For example, if the internuclear axis of the molecule is the z-axis, I_{zz} is zero in the linear molecule. For the linear case Watson (1970) derived the Hamiltonian

$$H = \tfrac{1}{2}\mu[(\Pi_x - \pi_x)^2 + (\Pi_y - \pi_y)^2] + \sum_{k=1}^{3N-5} \left(-\frac{h^2}{8\pi^2} \frac{\partial^2}{\partial Q_k^2} \right) + V(Q), \tag{4.61}$$

where

$$\mu = (I')^{-1} \tag{4.62}$$

with

$$I' = (I'')^2 (I^0)^{-1}. \tag{4.63}$$

I^0 is the moment of inertia for the equilibrium configuration

$$I^0 = \sum_i m_i (r^0_{z_i})^2 \tag{4.64}$$

and

$$I'' = I^0 + \sum_i \sum_k m_i^{1/2} r^0_{z_i} l_{z_i, k} Q_k. \tag{4.65}$$

In this case I'' will never be zero and no problems will be encountered with regard to divergence.

The derivation and detailed discussion of these Hamiltonians are beyond the scope of this book. For further discussion the reader is referred to Carney *et al.* (1978), Kroto (1975), Bunker (1979) and the original papers by Watson (1968, 1970). It is not possible to solve the Schrödinger equation exactly with the Hamiltonians of equations 4.55 and 4.61 so approximate methods must be used to derive the energy levels. Perturbation theory has been commonly used for spectroscopic problems but recently variational procedures have been devised for the calculation of vibration–rotation energy levels. More recently a scattering theory method has been used (Kidd *et al.* 1981).

4.2.2. Perturbation Theory

In perturbation theory, the Hamiltonian is expressed as the sum of a zero-order Hamiltonian H^0, for which exact solutions exist, and a perturbation $H^{(1)}$

$$H = H^0 + \lambda H^{(1)}, \tag{4.66}$$

where λ is an expansion parameter which is usually used for book-keeping purposes and then set to unity. Care must be taken not to confuse the perturbation parameter λ with λ_k for normal mode k as defined in equation 4.23. The wavefunctions ψ_i and energies E_i are also expanded in terms of the wavefunctions $\psi_i^{(0)}$ and energies $E_i^{(0)}$ for H^0, thus:

$$\psi_i = \psi_i^{(0)} + \lambda \psi_i^{(1)} + \lambda^2 \psi_i^{(2)} + \ldots, \tag{4.67}$$

$$E_i = E_i^{(0)} + \lambda E_i^{(1)} + \lambda^2 E_i^{(2)} + \ldots, \tag{4.68}$$

where $\psi_i^{(1)}$ and $E_i^{(1)}$ are the first-order corrections. For non-degenerate systems it can be readily shown (see, for example, Pauling and Wilson (1935)) that the first-order correction to the energy can be written

$$E_i^{(1)} = \int \psi_i^{(0)} H^{(1)} \psi_i^{(0)} \, d\tau. \tag{4.69}$$

If $\psi_i^{(1)}$ is expanded in terms of the complete set of the solutions $\psi_i^{(0)}$ of the zero-order Hamiltonian H^0

$$\psi_i^{(1)} = \sum_k a_{ik}\psi_k^{(0)} \tag{4.70}$$

then the second-order correction to the energy is given by

$$E_i^{(2)} = \sum_{j \neq i} \frac{H'_{ij}H'_{ij}}{E_i^0 - E_j^0}, \tag{4.71}$$

where

$$H'_{ij} = \int \psi_i^{(0)} H^{(1)} \psi_j^{(0)} \, d\tau. \tag{4.72}$$

In order to apply perturbation theory it is thus necessary to express the Hamiltonian in the form of equation 4.66. The zero-order Hamiltonian is the harmonic oscillator Hamiltonian

$$H^0 = \sum_{k=1}^{3N-6} \left(-\frac{h^2}{8\pi^2} \frac{\partial^2}{\partial Q_k^2} + \tfrac{1}{2}\lambda_k Q_k^2 \right), \tag{4.73}$$

where λ_k is related to the harmonic vibration wavenumber ω_k by $\lambda_k = 4\pi^2 c^2 \omega_k^2$. It is convenient to introduce P_k, the conjugate momentum to Q_k, defined as

$$P_k = -\frac{ih}{2\pi} \frac{\partial}{\partial Q_k}. \tag{4.74}$$

Thus

$$H^0 = \sum_{k=1}^{3N-6} \tfrac{1}{2}(P_k^2 + \lambda_k Q_k^2). \tag{4.75}$$

This can be written as

$$H^0 = \sum_{k=1}^{3N-6} \tfrac{1}{2}hc\omega_k(p_k^2 + q_k^2), \tag{4.76}$$

where p_k and q_k are dimensionless quantities defined by

$$q_k = \gamma_k^{1/2} Q_k \tag{4.77}$$

and

$$p_k = \frac{2\pi P_k}{\gamma_k^{1/2} h} \tag{4.78}$$

with

$$\gamma_k = \frac{2\pi \lambda_k^{1/2}}{h} = \frac{4\pi^2 c\omega_k}{h}, \tag{4.79}$$

where ω_k is the harmonic wavenumber for normal mode k. The elements of the modified inertia tensor $\mu_{\alpha\beta}$ can be expanded in powers of q_r thus

$$\mu_{\alpha\beta} = \mu^0_{\alpha\beta} + \sum_r \mu^{(r)}_{\alpha\beta} q_r + \tfrac{1}{2} \sum_r \sum_s \mu^{(r,s)}_{\alpha\beta} q_r q_s + \dots . \tag{4.80}$$

If the molecule-fixed axes are chosen to be the principal rotation axes for the equilibrium configuration

$$\mu^0_{\alpha\beta} = \frac{1}{I_\alpha} \delta_{\alpha\beta}, \tag{4.81}$$

where I_α is the equilibrium moment of inertia with respect to axis α and $\delta_{\alpha\beta}$ is the Kronecker delta ($= 1$ for $\alpha = \beta$ and 0 for $\alpha \neq \beta$),

$$\mu^{(r)}_{\alpha\beta} = \frac{\partial \mu_{\alpha\beta}}{\partial q_r} = -\frac{a_r^{(\alpha\beta)}}{\gamma_r^{1/2} I_\alpha I_\beta} \tag{4.82}$$

and

$$\mu^{(r,s)}_{\alpha\beta} = \frac{\partial^2 \mu_{\alpha\beta}}{\partial q_r \partial q_s} = \sum_\xi \frac{3(a_r^{(\alpha\xi)} a_s^{(\beta\xi)} + a_r^{(\beta\xi)} a_s^{(\alpha\xi)})}{4 \gamma_r^{1/2} \gamma_s^{1/2} I_\alpha I_\beta I_\xi}, \tag{4.83}$$

where $a_r^{(\alpha\beta)} = \partial I_{\alpha\beta} / \partial q_r$. The vibrational angular momentum π_α can be written in terms of p_r and q_r as

$$\pi_\alpha = \frac{h}{2\pi} \sum_k \sum_l \zeta_{kl}^{(\alpha)} q_k p_l \left(\frac{\omega_l}{\omega_k} \right)^{1/2}. \tag{4.84}$$

With the aid of equation 4.80 and the related formulae, the Watson Hamiltonian for a non-linear molecule can be expanded in terms of p_r, q_r and J_α, the angular momentum operator ($\Pi_\alpha = (h/2\pi)J_\alpha$). It is possible, in principle, to follow a conventional perturbation-theory approach using the eigenfunctions of the zero-order Hamiltonian (equation 4.75) in equations 4.67 and 4.69. These eigenfunctions will be products of harmonic oscillator wavefunctions and the appropriate rotational wavefunctions. The usual approach is to obtain an effective Hamiltonian for each vibrational state using perturbation theory with respect to the harmonic oscillator wavefunctions which are the eigenfunctions of the Hamiltonian operator

$$H^0 = \sum_{k=1}^{3N-6} \tfrac{1}{2} hc\omega_k (p_k^2 + q_k^2). \tag{4.85}$$

On expansion the Watson Hamiltonian yields an expression for H/hc which can be conveniently written in tabular form (table 4.1) to make clear the dependence of the various terms on powers of J and q (or p).

Table 4.1. Terms in rovibrational Hamiltonian, H/hc, arranged by order of magnitude and power of J (reproduced from Mills 1972).

Order of magnitude	Terms in H/hc involving J^0	J^1	J^2
$\kappa^0\nu_{\text{vib}}$	$\sum_r \frac{1}{2}\omega_r(p_r^2+q_r^2)$	–	–
$\kappa^1\nu_{\text{vib}}$	$+\sum_{rst}\frac{1}{6}\phi_{rst}q_rq_sq_t$	–	–
$\kappa^2\nu_{\text{vib}}$	$+\sum_{rstu}\frac{1}{24}\phi_{rstu}q_rq_sq_tq_u + \sum_\alpha B_e^{(\alpha)}[\,j_\alpha^2$	$-2j_\alpha J_\alpha$	$+J_\alpha^2]$
$\kappa^3\nu_{\text{vib}}$	$+$ quintic anharmonic $+$		
	$\sum_{\alpha\beta r}\frac{h^2}{8\pi^2c}\mu_{\alpha\beta}^{(r)}q_r[\,j_\alpha j_\beta$	$-(j_\alpha J_\beta+j_\beta J_\alpha)$	$+J_\alpha J_\beta]$
$\kappa^4\nu_{\text{vib}}$	$+$ sextic anharmonic $+$		
	$\sum_{\alpha\beta rs}\frac{h^2}{8\pi^2c}\mu_{\alpha\beta}^{(r,s)}q_rq_s[\,j_\alpha j_\beta$	$-(j_\alpha J_\beta+j_\beta J_\alpha)$	$+J_\alpha J_\beta]$

In this expansion j_α is the operator for the vibrational angular momentum $(\pi_\alpha = (h/2\pi)j_\alpha)$.

Expressions for the spectroscopic constants used in the analysis of the spectrum are obtained from the application of perturbation theory with harmonic oscillator wavefunctions. In a given vibrational state v, rotational energy levels are obtained by solving the rotational Schrödinger equation for that rotational level. In the general case of an asymmetric rotor, the equation is of the form

$$\frac{H_{\text{rot}}}{hc} = A_v J_a^2 + B_v J_b^2 + C_v J_c^2 + \frac{1}{4}\sum_\alpha\sum_\beta (\tau'_{\alpha\alpha\beta\beta})_v J_\alpha^2 J_\beta^2, \qquad (4.86)$$

where A_v, B_v and C_v are the effective rotational constants for the principal axes a, b and c and $(\tau'_{\alpha\alpha\beta\beta})_v$ represents the centrifugal distortion constants. Both the rotational constants and the centrifugal distortion constants depend on the particular vibrational level. The vibrational dependence of a rotational constant, e.g. B_v, is generally expressed in the form

$$B_v = B_e - \sum_k \alpha_k^B(v_k + \tfrac{1}{2}) + \sum_k\sum_{l\geq k}\gamma_{kl}^B(v_k + \tfrac{1}{2})(v_l + \tfrac{1}{2}), \qquad (4.87)$$

where the summation runs over all normal modes. If equation 4.87 for B_v is substituted into equation 4.86 there will be terms in the rotational Hamiltonian of the form $-\alpha_k^B(v_k + \tfrac{1}{2})J_b^2$ representing the vibration–rotation interaction between normal mode Q_k and rotation about the axis b. In order to calculate such a term by perturbation theory we have to identify all the contributions to it. This process is simplified by Watson's classification of the terms in table 4.1 according to the powers of (q_r, p_r) and J_α appearing in the various terms. If each term is expressed by $h_{m,n}$ where m is the power of (q_r, p_r) and n is the power of J_α, we obtain the table 4.2.

Table 4.2. Terms in the rovibrational Hamiltonian classified by order of magnitude and powers of (q_r, p_r) and J_α (reproduced from Mills 1972).

Order of magnitude	Powers of (q_r, p_r) and J_α		
	J^0	J^1	J^2
ν_{vib}	$h_{2,0}$	–	–
$\kappa\nu_{vib}$	$+h_{3,0}$	–	–
$\kappa^2\nu_{vib}$	$+h_{4,0}$	$+h_{2,1}$	$+h_{0,2}$
$\kappa^3\nu_{vib}$	$+h_{5,0}$	$+h_{3,1}$	$+h_{1,2}$
$\kappa^4\nu_{vib}$	$+h_{6,0}$	$+h_{4,1}$	$+h_{2,2}$

The terms are also classified according to powers of κ, the perturbation expansion parameter used in the Born–Oppenheimer expansion. $\kappa = (m/M)^{1/4}$, where m is the electronic mass and M is a typical nuclear mass. Using this parameter it can be shown that rotational (E_r), vibrational (E_v) and electronic (E_e) energies are of the order $E_r \approx \kappa^2 E_v$ and $E_v \approx \kappa^2 E_e$.

If, in a second-order perturbation calculation, $H^{(1)}$ contains two terms, say $h_{k,l}$ and $h_{m,n}$, the energy expression will contain terms such as

$$\sum_{v''} \sum_{R''} \frac{\int \psi_v \psi_R h_{k,l} \psi_{v''} \psi_{R''} \, d\tau \int \psi_{v''} \psi_{R''} h_{m,n} \psi_v \psi_{R'} \, d\tau}{E_{vR} - E_{v''R''}}. \qquad (4.88)$$

Since the rotation energies are very small compared with the vibration energies, the denominator can be written as $\Delta E_{vib} = E_v - E_{v''}$. The integrals are expressed as products of an integral over rotational functions and an integral over vibrational functions. Use of the properties of the angular momentum operators J and of the operators q and p enables one to reduce the product of integrals in the numerator of equation 4.88 to an integral of the form

$$\int \psi_v h^{eff} \psi_v \, d\tau,$$

where h^{eff} is an effective Hamiltonian for vibrational level v. It can be shown that one only obtains contributions from effective Hamiltonians having the structure $h^{eff}_{k+m-2, l+n}$ and $h^{eff}_{k+m, l+n-1}$ in the classification scheme of table 4.2.

If we are interested in the vibration–rotation interaction term we see that $-\alpha_k^B(v + \frac{1}{2})J(J+1)$ is an energy term arising from an effective Hamiltonian of the type $h^{eff}_{2,2}$ because we require it to contain the J^2 operator and in order to obtain the vibrational eigenvalue $v + \frac{1}{2}$ we have to average a vibrational operator of order 2, e.g.

$$\int \psi_v (\tfrac{1}{2}q_r^2 + \tfrac{1}{2}p_r^2) \psi_v \, d\tau = v + \tfrac{1}{2}.$$

There are three ways in which terms of this type can arise. In first-order perturbation theory, where the energy is given by an integral of the type in equation 4.69, we require $h_{2,2}^{\text{eff}} = h_{2,2}$. Reference to table 4.1 shows that this is a term of the type $q_r q_s J_\alpha J_\beta$ arising from the quadratic dependence of $\mu_{\alpha\beta}$ (or B) on the normal coordinates. Two types of term contribute in second-order perturbation theory, namely $h_{2,2}^{\text{eff}} = (h_{2,1} \times h_{2,1})/\Delta E_{\text{vib}}$ and $(h_{3,0} \times h_{1,2})/\Delta E_{\text{vib}}$. The first arises from terms of the type $j_\alpha J_\alpha$ corresponding to a second-order Coriolis interaction between the modes with vibration wavenumbers ω_r and ω_s. The second term arises from cubic terms $f_{rrs} q_r^2 q_s$ in the potential energy and terms of the type $q_s J_\alpha J_\beta$. The interpretation of this contribution is that because of the anharmonicity, a mean square displacement of q_r causes a mean first-power displacement of q_s with a consequent change in the effective rotation constant. By using standard expressions for integrals with respect to the various operators, a formula for the vibration–rotation interaction can be derived.

More detailed discussions of the calculation of effective rotational Hamiltonians by the methods outline above are give by Mills (1972), Allen and Cross (1963) and Kroto (1975). The use of perturbation theory enables one to calculate the various interaction constants from a knowledge of the atomic masses, the equilibrium structure and the force field. Vibration–rotation spectra can then be calculated by the use of the selection rules and intensity formulae.

4.2.3. The Variational Method

The discussion in the previous section was limited to second-order perturbation theory. It is possible to extend the method to fourth-order but the treatment becomes increasingly complicated as one goes to higher order in perturbation theory. Difficulties arise when higher order terms in the potential are important and when there are degeneracies or near-degeneracies in the energy levels. Another disadvantage of perturbation theory is that it is much more difficult to obtain the wavefunctions than the energies. It is therefore not easy to use the results in the calculation of transition probabilities which requires the evaluation of integrals involving the wavefunctions for the ground and excited states. In view of these difficulties, attention has been directed in recent years to the use of variational procedures for the calculation of vibration–rotation energy levels in small molecules.

In this approach the wavefunction Ψ_{JM} for a given vibration–rotation state is written as a linear combination of basis functions $\psi_n R_{JKM}$

$$\Psi_{JM} = \sum_n \sum_K C_{nK}^{(JM)} \psi_n R_{JKM}, \tag{4.89}$$

where ψ_n is a suitable vibrational wavefunction, R_{JKM} is a symmetric top wavefunction and $C_{nK}^{(JM)}$ is a variational parameter. The energies and wavefunctions for the levels of interest are then calculated by applying the condition that the energy

$$\int \psi_{JM} H \psi_{JM} \, d\tau$$

is stationary with respect to the variational parameters $C_{nK}^{(JM)}$. This leads to a set of homogeneous simultaneous equations, the secular equations, which involve the integrals

$$\int \psi_{n'} R_{J'K'M'} H \psi_n R_{JKM} \, d\tau,$$

where H is the vibration–rotation Hamiltonian. The choice of Hamiltonian operator is discussed in detail by Carney *et al.* (1978). Several calculations have used the Watson Hamiltonian (for non-linear or linear molecules as appropriate) expressed in terms of normal coordinates Q_i. The vibrational part of the basis functions used in the expansion of equation 4.89 is written, for a triatomic molecule, as a product of harmonic oscillator functions

$$\psi_n = H_{n_1}(q_1) H_{n_2}(q_2) H_{n_3}(q_3) \exp[-\tfrac{1}{2}(q_1^2 + q_2^2 + q_3^2)], \qquad (4.90)$$

where $H_{n_i}(q_i)$ are Hermite polynomials. The basis set can be systematically increased in size by taking all possible combinations of n_1, n_2, n_3 such that $n_1 + n_2 + n_3 \leqslant N$ where N is an integer. In H_2O, for example, choice of $N = 6$ gives convergence to within $0.5 \, \text{cm}^{-1}$ for the (000), (010), (100) and (001) levels and to within $3 \, \text{cm}^{-1}$ for the (020) level (where $(n_1 n_2 n_3)$ represents the number of quanta in the symmetric stretching, bending and asymmetric stretching modes, respectively). This method seems to work well for the lower vibrational levels of molecules such as SO_2 which are strongly bent and for linear molecules such as OCS. Provided that a suitable potential function is chosen, reasonable agreement is obtained with experimental band origins. However, problems do arise in the calculation of higher vibrational levels and spurious results will be obtained when the energy levels are comparable with the barrier height. These problems are particularly acute in molecules which have a relatively low barrier to linearity. An example is CH_2^+ for which the barrier height is of the order of $900–1000 \, \text{cm}^{-1}$.

For such molecules it is more appropriate to choose a Hamiltonian expressed, in the case of a triatomic molecule, in terms of two bond lengths and the included angle. The use of harmonic oscillator functions in the expansion in equation 4.89 would also seem to be somewhat inappropriate for higher vibrational levels. More rapid convergence can be obtained by using, for the stretching coordinates, Morse oscillator wavefunctions obtained from the solution of the Schrödinger equation for the Morse potential (equation 2.16). Details are given by Carter and Handy (1982). When $J = 0$, the wavefunction for the bending coordinate can be expressed in terms of Legendre polynomials. However, complications do arise and it is necessary to perform a separate angular calculation to choose the angular basis. Comparison with the earlier methods shows that this approach is superior for linear or quasi-linear molecules but for strongly bent molecules it is necessary to use a much larger basis of angular functions.

4.3. Vibrational Energy Levels for Polyatomic Molecules

4.3.1. *Vibrational Energy Levels for an Anharmonic Potential*

The potential energy function for a polyatomic molecule is of the general form of equation 4.2. If there are no degenerate vibrations, the vibrational spectrum can be fitted to an empirical energy expression of the form

$$G(v_1, v_2, \ldots) = \sum_i \omega_i(v_i + \tfrac{1}{2}) + \sum_i \sum_{j \geq i} x_{ij}(v_i + \tfrac{1}{2})(v_j + \tfrac{1}{2}), \qquad (4.91)$$

with $x_{ij} \ll \omega_i$. The terms $\omega_i(v_i + \tfrac{1}{2})$ represent the harmonic contributions and the remaining terms arise from the anharmonicity. This equation is a generalization of the linear and quadratic terms in equation 2.14 for a diatomic molecule. For molecules having doubly degenerate vibrational modes, i.e. linear and symmetric top molecules, equation 4.91 is modified to

$$G(v_1, v_2, \ldots) = \sum_i \omega_i(v_i + \tfrac{1}{2}d_i)$$

$$+ \sum_i \sum_{j \geq i} x_{ij}(v_i + \tfrac{1}{2}d_i)(v_j + \tfrac{1}{2}d_i) + \sum_i \sum_{j \geq i} g_{ij}l_i l_j, \quad (4.92)$$

where d_i is the degeneracy of mode i. This equation contains an additional term $\sum g_{ij}l_i l_j$ to account for the angular momentum arising from the excitation of degenerate modes by several quanta. For example, in a linear XYX molecule, if there is one quantum in each of the degenerate bending modes, with a phase difference of 90°, there is a resultant rotational motion about the internuclear axis. The quantum number l_i, which can be either positive or negative, for mode i is related to the vibrational quantum number v_i by

$$|l_i| = v_i, v_i - 2, v_i - 4, \ldots, 1 \text{ or } 0. \qquad (4.93)$$

The magnitude of the vibrational angular momentum $|\pi_z|$ is given by

$$\frac{h}{2\pi} \left| \sum_i l_i \right|.$$

However, there are several types of molecule for which the potential function of equation 4.2 and the energy level expressions of equations 4.91 and 4.92 are not appropriate. Many potential surfaces have more than one minimum, as discussed in chapter 3 for the cases of HCN and O_3. In the case of HCN and HNC the two minima have different energies and shapes and each one can be considered separately. The molecule cannot pass from one to the other in the course of a vibration. There are other molecules for which the potential surfaces have several identical minima connected by a vibrational mode of the molecule. We review briefly some of these cases in the following sections.

4.3.2. Inversion Doubling

A pyramidal molecule such as NH_3 has two identical minima connected by an 'umbrella' vibrational mode ν_2 in which the nitrogen atom passes through the plane defined by the three hydrogen atoms. The potential energy curve for this vibrational mode is shown in figure 4.2 and consists of two identical minima separated by a potential barrier.

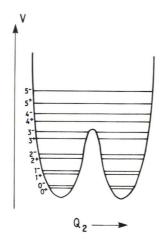

Figure 4.2. Potential energy curve for 'umbrella' vibrational mode ν_2 of NH_3 showing inversion doubling of vibrational levels.

If the barrier were very high and very wide, one could regard the two branches as being represented by two harmonic potential functions. Each would have a set of equally spaced energy levels according to equation 2.8. However, for a molecule such as NH_3, the height and width of the barrier are such that quantum mechanical tunnelling occurs, resulting in an interaction between the energy levels and wavefunctions for the two minima. If the wavefunctions for the left- and right-hand minima are denoted $\psi_v^{(l)}$ and $\psi_v^{(r)}$

respectively, for a given quantum number v, in order to account for the interaction between the levels we take linear combinations of $\psi_v^{(l)}$ and $\psi_v^{(r)}$. The result is simply the symmetric and antisymmetric combinations

$$\psi_s = \psi_v^{(l)} + \psi_v^{(r)}, \qquad \psi_a = \psi_v^{(l)} - \psi_v^{(r)}. \tag{4.94}$$

These two combinations differ in energy with ψ_s being the lower. Thus each of the levels splits into two and the splitting increases as one approaches the top of the barrier. The levels are labelled $0^+, 0^-, 1^+, 1^-$ etc. in order of increasing energy. The energy separation between the 0^+ and 0^- levels in NH_3 is such that the wavenumber for the transition $(0.793\ cm^{-1})$ occurs in the microwave region of the spectrum. The potential function with respect to normal mode Q_2 of NH_3 can be represented by

$$V = \tfrac{1}{2}aQ^2 + b\exp(-aQ^2). \tag{4.95}$$

4.3.3. Ring Puckering Vibrations

Many cyclic molecules have low frequency vibrational modes corresponding to out-of-plane puckering of the ring. For example, the equilibrium structure of cyclobutane is non-planar and in the puckering vibration the molecule vibrates between the two equivalent potential minima. This motion is very anharmonic and it is found that the vibrational energy levels can be fitted to a potential of the form

$$V = -aQ^2 + bQ^4. \tag{4.96}$$

The form of the potential is illustrated in figure 4.3. Levels below the potential barrier may be split because of tunnelling.

4.3.4. Torsional Motion

In a molecule such as C_2H_6, rotation of one methyl group through 120° or 240° about the carbon–carbon bond transforms the molecule into an identical configuration. If one starts with the molecule at a potential minimum, there must be identical minima on rotation through 120° or 240° and the potential function with respect to the rotation can be represented by figure 4.4. The functional form of this potential can be written in terms of the angle of rotation ϕ as

$$V(\phi) = \tfrac{1}{2} \sum_k V_k(1 - \cos k\phi). \tag{4.97}$$

If there are n identical minima, then only terms with $k = pn$, where p is an

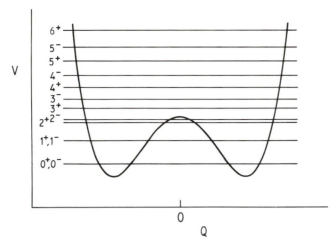

Figure 4.3. Potential energy curve for ring-puckering mode of cyclobutane. (Adapted from J. M. R. Stone and I. M. Mills, *Mol. Phys.*, **18** 631 (1970).)

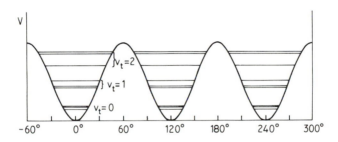

Figure 4.4. Potential energy curve as a function of angle of torsion for a C_2H_6-like molecule. (Adapted from Herzberg 1945.)

integer, are allowed by symmetry and the term with $k = n$ is dominant. The molecule undergoes torsional motion if it does not have sufficient energy to surmount the barrier. Free rotation occurs when the energy is in excess of the barrier height. There is no analytical expression for the energy levels for such a potential. A schematic energy level scheme is shown in figure 4.4. Near the potential minimum tunnelling is unimportant and the energy levels get closer with increasing v as in the anharmonic oscillator. Nearer the top of the barrier tunnelling does become important resulting in the splitting of each level into n levels where n is the number of equivalent minima. This splitting increases as the top of the barrier is approached.

4.4. Interaction of Vibrational Motion and Electronic Motion

4.4.1. Linear Molecules: The Renner–Teller Effect

In linear molecules in which the electronic state is degenerate (i.e., for states other than Σ states) complications arise in bending modes from the interaction between the electronic motion and the vibrational motion. In such situations the Born–Oppenheimer separation is no longer valid and one has to determine the vibronic energy levels. The degeneracy of such electronic states (i.e., Π, Δ states etc.) is twofold, but when the molecule is bent this degeneracy is removed and the potential curve in terms of the bending coordinates splits into two curves as shown in figure 4.5(a) for a Π state.

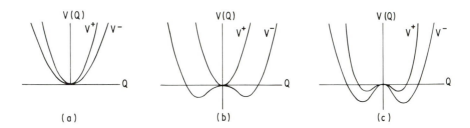

Figure 4.5. Splitting of the potential function on bending a linear molecule (a) for a Π state; (b) for the case where one component is linear and the other bent; (c) for the case where both components have bent equilibrium geometries but are degenerate when linear.

The potential function for a bending mode must be symmetric with respect to the equilibrium configuration and can thus be represented by a function of the form

$$V = aQ^2 + bQ^4. \tag{4.98}$$

The splitting between the two curves V^+ and V^- obtained on bending the molecule must also be of the same form

$$V^+ - V^- = \alpha Q^2 + \beta Q^4. \tag{4.99}$$

If the term α is so large that $\frac{1}{2}\alpha > a$, the V^- curve will have two minima with a maximum at $Q = 0$, as illustrated in figure 4.5(b). This corresponds to the situation where one component of the state is linear and the other bent. A third situation arises when the equilibrium configurations of both components are bent but the states become degenerate for linear geometries. This is shown in figure 4.5(c). For Δ states α is zero and the splitting only becomes appreciable for large distortions.

In this situation the Born–Oppenheimer separation breaks down and we are not able to consider electronic and vibrational motion separately. We can no longer write the wavefunction for the energy levels simply as a product of an electronic wavefunction and a vibrational wavefunction nor is the energy the sum of the two individual energies. The breakdown of the Born–Oppenheimer separation occurs because the two potential surfaces are in contact and the molecule may change from one surface to the other. The energy levels cannot be associated with just one surface. The quantum numbers Λ (for the component of the orbital angular momentum along the internuclear axis) and l (for the vibrational angular momentum as defined in equation 4.93) are no longer good quantum numbers. Instead we define a quantum number K for the resultant vibronic angular momentum by the relationship

$$K = \Lambda + l. \tag{4.100}$$

In a Π state $|\Lambda| = 1$ and $|l|$ has the values 0 (for $v = 0$); 1 (for $v = 1$); 0, 2 (for $v = 2$); 1, 3 (for $v = 3$) etc. Thus the allowed values of $|K|$ are 1 (for $v = 0$); 0, 0, 2 (for $v = 1$); 1, 1, 3 (for $v = 2$); 0, 0, 2, 2, 4 (for $v = 3$), etc. Levels are labelled Σ, Π, Δ, Φ, etc. for values of $|K|$ equal to 0, 1, 2, 3, etc. When the interaction between the vibrational and electronic motions is taken into account distinct vibronic energy levels are obtained for each of the values of $|K|$. Figure 4.6 shows the correlation between the vibronic levels and the vibrational levels labelled with v and $|l|$. The value $K = 0$ always arises twice from combinations such as $\Lambda = +1, l = -1$ and $\Lambda = -1, l = +1$ and we obtain Σ^+ and Σ^- levels. In molecules where the electronic spin is strongly coupled to the orbital angular momentum, we have to use the quantum number Ω for the total electronic angular momentum instead of Λ in equation 4.100 to give a quantum number P for the resultant vibronic angular momentum including spin

$$P = \Omega + l. \tag{4.101}$$

If the quartic terms in equation 4.99 are neglected, the splitting between the Σ^+ and Σ^- levels, for which $K = 0$, in a $^1\Pi$ state is given by the formula

$$G^{\pm}(v, 0) = \omega(1 \pm \epsilon)^{\frac{1}{2}}(v + 1) \tag{4.102}$$

where the $\pm$ sign refers to the Σ^+ and Σ^- states respectively, ω is the harmonic vibration wavenumber and v is the vibrational quantum number for the bending mode in question. v can take the values 1, 3, 5, etc. since even values of v do not give rise to Σ^+, Σ^- states. If $K \neq 0$, the energy of the lowest single vibronic level of a given species ($v = |K| - 1$) is given by

$$G(v, K) = \omega[(v + 1) - \tfrac{1}{8}\epsilon^2 |K|(|K| + 1)], \tag{4.103}$$

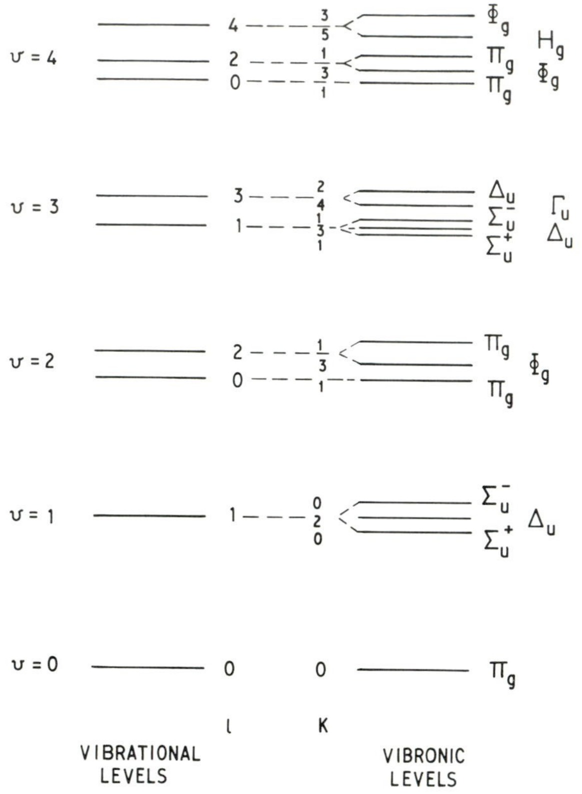

Figure 4.6. Correlation between vibronic levels and vibrational levels for a linear molecule.

whereas for the other levels $(v > |K| - 1)$

$$G^{\pm}(v, K) = \omega(1 - \tfrac{1}{8}\epsilon^2)(v + 1) \pm \tfrac{1}{2}\omega\epsilon[(v + 1)^2 - K^2]^{\frac{1}{2}}. \qquad (4.104)$$

The Renner parameter $\epsilon = \alpha/2a$ is a measure of the strength of the vibronic interaction. These energy expressions should be used in place of the term $\omega_2(v_2 + 1) + g_{22}l_2^2$ for bending mode v_2 in equation 4.92. The effects of anharmonicity have not been included in this treatment. The vibronic splitting of a level with given v and $|l|$ values is known as the Renner–Teller splitting.

A more complete discussion of the Renner–Teller effect for both singlet and triplet states is given by Longuet-Higgins (1961), Herzberg (1966), Duxbury (1975) and Jungen and Merer (1976).

4.4.2. *Non-Linear Molecules: The Jahn–Teller Effect*

Vibronic effects are also important in electronically degenerate states of symmetric top and spherical top molecules, which have a symmetry axis of order larger than 2. For linear molecules the bending motion removes the degeneracy but in many cases the minima in the potential curves occur for the linear configuration. The same situation may occur for some non-totally symmetric vibrational modes in symmetric top or spherical top molecules, but the important difference is that the Jahn–Teller theorem states that there is always at least one non-totally symmetric normal coordinate for which the splitting results in two or more separate potential minima for non-zero values of the normal coordinate. This is illustrated in figure 4.7 where it can be seen that for $Q = 0$ the two potential curves cross at a non-zero angle in contrast to the Renner–Teller effect in which the potential curves have zero slope at the point of contact for zero displacement.

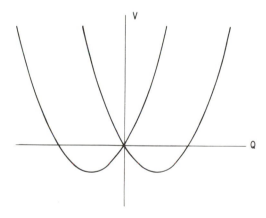

Figure 4.7. Potential function for a non-totally symmetric vibrational mode of a symmetric top or spherical top molecule showing the Jahn–Teller effect.

This phenomenon is well known in transition metal chemistry where an octahedral structure for an ion containing, say, nine d-electrons would have the electronically degenerate configuration $t_{2g}^6 e_g^3$. Because of the Jahn–Teller theorem there will be a structure of lower symmetry which can be obtained by distorting the regular octahedral structure along one of the normal coordinates.

Figure 4.7 is drawn for a non-degenerate vibrational mode and the result is two potential minima. Let us consider a molecule with a threefold rotation axis such as CH_3I in a doubly degenerate electronic ^1E state in which a doubly degenerate vibrational mode is excited. The variation of potential energy with the second component of the vibration will also be represented by a figure like 4.7 but rotated through 90°. In the simplest approximation, the coupling between the two surfaces is taken as a linear function of the normal coordinate resulting in an energy lowering proportional to $(Q_a^2 + Q_b^2)^{1/2}$. At this level, the complete potential surface is obtained by rotating figure 4.7 about the *V*-axis. Inclusion of quadratic terms in the coupling results in a surface which has three identical minima with respect to rotation about the *V*-axis. This is illustrated in figure 4.8.

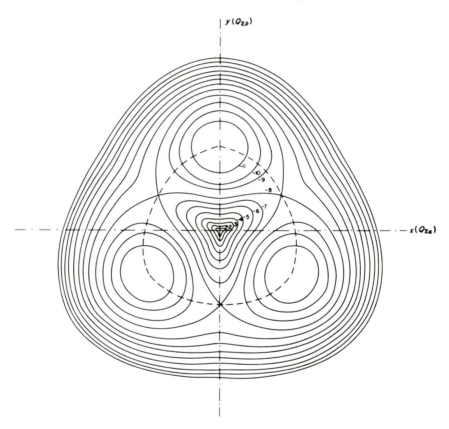

Figure 4.8. Potential surface for a doubly degenerate vibrational mode of a C_{3v} (or D_{3h}) molecule in a degenerate electronic state. Quadratic terms have been included in the vibronic interaction. (Reproduced from Herzberg 1966.)

The occurrence of several minima in such molecules due to vibronic interactions is known as the static Jahn–Teller effect. If the interaction is large then perhaps one should regard the molecule as being unsymmetrical. However, we are mainly concerned with cases where the interaction is relatively small.

We now consider the effect of this interaction on the vibrational energy levels or the dynamic Jahn–Teller effect. In the absence of vibronic interactions, for each value of v there will be a number of sub-levels, some of which may be degenerate, characterized by the vibrational quantum number l. Each of the sub-levels can be classified according to symmetry (see Herzberg 1945). Thus if we have a molecule such as NH_3, which has C_{3v} symmetry, we get the levels shown in figure 4.9 for various numbers of quanta in the doubly degenerate vibrational mode of symmetry E.

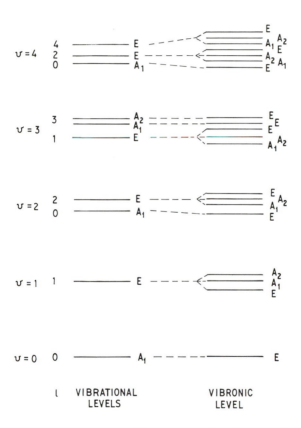

Figure 4.9. Vibronic levels for molecule of C_{3v} symmetry in a degenerate E electronic state for a doubly degenerate vibrational mode of symmetry E.

If the molecule is in an electronically degenerate state, say ^1E, then we have to determine the vibronic symmetry species by taking the direct product of the irreducible representation of the electronic wavefunction with the irreducible representation of the vibrational wavefunction. The results of this are shown in figure 4.9. The vibronic interaction results in splitting into as many different vibronic levels as there are vibronic species for a given value of v. These splittings are known as Jahn–Teller splittings. Degenerate vibronic levels, say of E character, are not split by vibronic interactions.

In the simplest treatment in which only terms linear in the normal coordinate are included in the interaction integrals, there is no splitting between levels of A_1 and A_2 symmetry but more complete treatments do give a splitting. For small vibronic interactions the energy levels are given by

$$G(v, l) = \omega(v + 1) \mp 2D\omega(l \pm 1), \qquad (4.105)$$

where $D\omega$ is the depth of the potential trough below the energy for zero distortion given on rotation of figure 4.7 about the V-axis. Only linear terms have been included in the interaction integrals and the expression is only valid for $D < 0.05$. Figure 4.10 shows the pattern of splittings obtained for a molecule of C_{3v} symmetry. More detailed treatments of the Jahn–Teller effect are given by Longuet-Higgins (1961), Herzberg (1966) and Englman (1972).

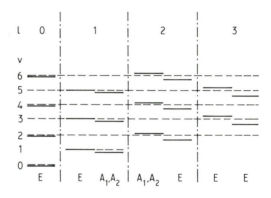

Figure 4.10. Jahn–Teller splittings for a molecule of C_{3v} symmetry. (Adapted from Herzberg 1966.)

Suggestions for Further Reading

G. Herzberg (1945) *Molecular Spectra and Molecular Structure*, Vol. II. *Infra-red and Raman Spectra of Polyatomic Molecules*, Van Nostrand, Princeton.

G. Herzberg (1966) *Molecular Spectra and Molecular Structure*, Vol. III. *Electronic Spectra and Electronic Structure of Polyatomic Molecules*, Van Nostrand, Princeton.

J. M. Hollas (1982) *High Resolution Spectroscopy*, Butterworths, London.

G. W. King (1964) *Spectroscopy and Molecular Structure,* Holt, Rinehart and Winston, New York.
I. M. Mills (1963) in *Infra-Red Spectroscopy and Molecular Structure,* ed. Mansel Davies, p. 166, Elsevier, Amsterdam.
L. A. Woodward (1972) *Introduction to the Theory of Molecuar Vibrations and Vibrational Spectroscopy,* Clarendon Press, Oxford.

References

H. C. Allen and P. C. Cross (1963) *Molecular Vib-Rotors,* Wiley, New York.
P. R. Bunker (1979) *Molecular Symmetry and Spectroscopy,* Academic Press, New York.
G. D. Carney, L. L. Sprandel and C. W. Kern (1978) *Adv. Chem. Phys.,* **37**, 305.
S. Carter and N. C. Handy (1982) *Mol. Phys.,* **47**, 1445.
J. L. Duncan (1975) in *Molecular Spectroscopy,* Vol. 3, ed. R. F. Barrow, D. A. Long and D. J. Millen, p. 104, The Chemical Society, London.
G. Duxbury (1975) in *Molecular Spectroscopy,* Vol. 3, ed. R. F. Barrow, D. A. Long and D. J. Millen, p. 497, The Chemical Society, London.
R. Englman (1972) *The Jahn–Teller Effect in Molecules and Crystals,* Wiley–Interscience, London.
P. Gans (1971) *Vibrating Molecules; An Introduction to the Interpretation of Infra-Red and Raman Spectra,* Chapman and Hall, London.
G. Herzberg (1966) *Molecular Spectra and Molecular Structure,* Vol. III. *Electronic Spectra and Electronic Structure of Polyatomic Molecules,* Van Nostrand, Princeton.
A. R. Hoy, I. M. Mills and G. Strey (1972) *Mol. Phys.,* **24**, 1265.
Ch. Jungen and A. J. Merer (1976) in *Molecular Spectroscopy: Modern Research,* Vol. II, ed. K. N. Rao, p. 127, Academic Press, New York.
I. F. Kidd, G. G. Balint-Kurti and M. Shapiro (1981) *Farad. Disc. Chem. Soc.,* **71**, 287.
H. W. Kroto (1975) *Molecular Rotation Spectra,* Wiley, London.
H. C. Longuet-Higgins (1961) *Advances in Spectroscopy,* **2**, 429.
I. M. Mills (1972) in *Molecular Spectroscopy: Modern Research,* Vol. I, ed. K. N. Rao and C. W. Mathews, p. 115, Academic Press, New York.
I. M. Mills (1974) in *Theoretical Chemistry,* Vol. 1, ed. R. N. Dixon, p. 110, The Chemical Society, London.
L. Pauling and E. B. Wilson (1935) *Introduction to Quantum Mechanics,* McGraw-Hill, New York.
T. Shimanouchi (1970) in *Physical Chemistry; An Advanced Treatise,* Vol. IV, ed. D. Henderson, p. 233, Academic Press, New York.
B. T. Sutcliffe (1980) in *Quantum Dynamics of Molecules,* ed. R. G. Woolley, p.1, Plenum Press, New York.
B. T. Sutcliffe (1982) in *Studies in Physical and Theoretical Chemistry 21, Current Aspects of Quantum Chemistry 1981,* ed. R. Carbó, p. 99, Elsevier, Amsterdam.

J. K. G. Watson (1968) *Mol. Phys.*, **15**, 479.

J. K. G. Watson (1970) *Mol. Phys.*, **19**, 465.

E. B. Wilson, J. C. Decius and P. C. Cross (1955) *Molcular Vibrations*, McGraw-Hill, New York.

L. A. Woodward (1972) *Introduction to the Theory of Molecular Vibrations and Vibrational Spectroscopy*, Clarendon Press, Oxford.

CHAPTER 5

theoretical calculation of potential energy surfaces

5.1. Introduction

In section 1.2 we defined the potential energy surface or potential energy function in terms of the Born–Oppenheimer separation. The nuclear energy levels are obtained by solving the nuclear Schrödinger equation

$$[T_{nuc}(\boldsymbol{R}) + V_{nn}(\boldsymbol{R}) + W(\boldsymbol{R})]\chi(\boldsymbol{R}) = E\chi(\boldsymbol{R}), \tag{5.1}$$

where $W(\boldsymbol{R})$ is the energy obtained by solving the electronic Schrödinger equation in the clamped nuclei approximation

$$[T_{el}(\boldsymbol{r}) + V_{ne}(\boldsymbol{R}, \boldsymbol{r}) + V_{ee}(\boldsymbol{r})]\Phi(\boldsymbol{R}, \boldsymbol{r}) = W(\boldsymbol{R})\Phi(\boldsymbol{R}, \boldsymbol{r}). \tag{5.2}$$

The potential energy surface is defined by the sum $W(\boldsymbol{R}) + V_{nn}(\boldsymbol{R})$. Thus a theoretical calculation of the potential energy surface involves the solution of equation 5.2 for a sufficiently large set of geometries. Since the electronic Schrödinger equation can only be solved exactly for one-electron systems, it is clearly necessary to make some approximations. There are many methods which can be used to calculate potential energy surfaces. These methods can be divided into four categories. We consider first *ab initio* methods in which a particular model is chosen for the electronic wavefunction and the calculation is then performed at a particular level without any further approximations. The success of an *ab initio* calculation depends on the appropriate choice of model and on the level at which the calculation is done. In some cases *ab initio* calculations can give very accurate results which are comparable with experimental data. On the other hand the wrong choice of a model can lead to results which are grossly in error and very misleading. Thus the label *ab initio* does not confer any stamp of authority on the calculation. One has to look critically at the method employed in order to assess the value of the calculations. Accurate potential energy surfaces can only be calculated by *ab initio* methods for relatively small molecules containing, in general, light atoms. For larger systems the calculations become prohibitive both in terms of computer time and storage requirements. Thus many attempts have been made to devise semi-theoretical methods which can reproduce the results of fully *ab initio* calculations for small systems and which can be used for larger molecules or for molecules containing heavy atoms. An alternative approach is the development of semi-empirical methods which are usually based on the formalism of *ab initio* methods. Several approximations, which are often very drastic, are made and in order to compensate for errors which may be introduced by these approximations some experimental parameters are used to ensure that the calculations yield reasonable results. The calculations usually aim to reproduce experimental data rather than mimic *ab initio* calculations. The fourth group of methods consists of purely empirical methods. These may involve the use of an empirical analytical function to fit experimental or *ab initio* data.

Many *ab initio*, semi-theoretical and semi-empirical methods are based on the molecular orbital method even though it is ill suited to the calculation of potential surfaces. We will begin our discussion of *ab initio* methods with a résumé of the molecular orbital method and then consider methods which will remedy its deficiencies. The valence bond method will be discussed briefly in section 5.2.3. We can only give an outline here of the main methods used for the calculation of potential surfaces. For a more detailed discussion of these methods and of their applicability the reader should consult more specialized books and review articles. Many-body perturbation theory and electron pair methods have been used to a limited extent for potential surface calculations but will not be discussed here. Comprehensive reviews of potential energy surface calculations have been given by Balint-Kurti (1975), Bader and Gangi (1975) and Hirst (1982). Books by Mulliken and Ermler (1977, 1981) discuss the application of *ab initio* methods to diatomic and polyatomic molecules. Bibliographies by Richards *et al.* (1971, 1974, 1978, 1981) and by Ohno and Morokuma (1982) document *ab initio* calculations up to the end of 1980 and are invaluable sources of information on calculations of potential energy surfaces.

5.2. *Ab Initio* Methods

5.2.1. *The Self-Consistent Field Molecular Orbital (SCF–MO) Method*

Much of our thinking about molecular electronic structure is in terms of molecular orbitals. Thus a discussion of the molecular orbital method is a useful starting point even though the method is not capable, in general, of giving a satisfactory description of a potential energy surface as one goes from a stable molecule to dissociation fragments or from a transition state to reactants or products. Reasons for this will be discussed below.

It is assumed that the reader has a knowledge of basic molecular orbital theory so our discussion adopts a more fundamental viewpoint than that usually taken in introductory texts and courses. The molecular orbital method is based on an independent particle model for the molecular electronic wavefunction. By this we mean that a form of wavefunction is chosen which would be appropriate if we could ignore the interelectronic repulsion (the term $V_{ee}(r)$ in equation 5.2). If this were the case then the electronic Schrödinger equation (5.2) for n electrons would be separable into a set of n one-electron Schrödinger equations which could be solved to obtain a set of one-electron functions (often called spin orbitals) $\phi_i(r)$, where $i = 1, 2, \ldots, n$. The total electronic wavefunction could then be written as a product of one-electron functions

$$\Phi(R, r) = \phi_1(r_1)\phi_2(r_2)\ldots\phi_n(r_n). \tag{5.3}$$

Each one-electron function $\phi_i(r_i)$ can be written as the product of a spatial function $\psi_i(r_i)$, which depends on the coordinates r_i of electron i, and a spin function α or β, depending on whether the spin is up or down. Each spatial orbital can hold two electrons — one with α-spin and one with β-spin. Assuming double occupancy and an even value for n, we can rewrite the electronic wavefunction $\Phi(R, r)$ as

$$\Phi(R, r) = \psi_1(r_1)\alpha\psi_1(r_2)\beta\psi_2(r_3)\alpha\ldots\psi_{n/2}(r_{n-1})\alpha\psi_{n/2}(r_n)\beta. \tag{5.4}$$

However, it is necessary to take into account the indistinguishability of the electrons by recognizing that we can permute the electron labels among the orbitals. At the same time we have to ensure that the wavefunction satisfies the Pauli principle by being antisymmetric with respect to the interchange of two electrons by taking the appropriate linear combination of all permutations. Both requirements can be satisfied by writing the wavefunction as a Slater determinant

$$\Phi(R, r) = \frac{1}{\sqrt{n!}} \begin{vmatrix} \psi_1(r_1)\alpha & \psi_1(r_1)\beta & \cdots & \psi_{n/2}(r_1)\beta \\ \psi_1(r_2)\alpha & \psi_1(r_2)\beta & \cdots & \psi_{n/2}(r_2)\beta \\ \vdots & \vdots & \ddots & \vdots \\ \psi_1(r_n)\alpha & \psi_1(r_n)\beta & \cdots & \psi_{n/2}(r_n)\beta \end{vmatrix}. \tag{5.5}$$

Interchange of two electrons is equivalent to interchanging the labels for the electronic coordinates, e.g. r_i with r_j resulting in a determinant differing from equation 5.5 by the interchange of the rows corresponding to r_i and r_j. This results in a change of sign, as required by the Pauli principle. The Slater determinant of equation 5.5 will be represented below by the more compact notation

$$\Phi = |\psi_1\alpha \; \psi_1\beta \; \psi_2\alpha \; \psi_2\beta \; \ldots \; \psi_{n/2}\alpha \; \psi_{n/2}\beta|. \tag{5.6}$$

The orbitals ψ_i are calculated by applying the variation method to minimize the energy corresponding to the wavefunction $\Phi(\boldsymbol{R}, \boldsymbol{r})$. This is given by the expectation value of the electronic Hamiltonian and is expressed as the integral

$$W_{\text{approx}} = \frac{\int \Phi(\boldsymbol{R}, r)^* H_{\text{el}}(\boldsymbol{R}, r) \Phi(\boldsymbol{R}, r) \, \mathrm{d}\tau}{\int \Phi(\boldsymbol{R}, r)^* \Phi(\boldsymbol{R}, r) \, \mathrm{d}\tau}. \tag{5.7}$$

The electronic Hamiltonian is given by

$$H_{\text{el}}(\boldsymbol{R}, r) = T_{\text{el}}(r) + V_{\text{ne}}(\boldsymbol{R}, r) + V_{\text{ee}}(r). \tag{5.8}$$

The integration is over the whole of space and $\mathrm{d}\tau$ represents the appropriate volume element. The variation theorem (see Pauling and Wilson 1935) states that W_{approx} is an upper bound to the exact electronic energy $W(\boldsymbol{R})$. Thus, having assumed a molecular wavefunction of the form of equation 5.5, we are seeking the 'best' set of molecular orbitals ψ_i by the criterion that W_{approx} given by equation 5.7 is a minimum, subject to the constraint that the orbitals ψ_i are orthogonal. This condition leads to the Hartree–Fock equations for the orbitals ψ_i. The derivation is rather complicated and will not be given here (see Parr 1963, Slater 1963). The Hartree–Fock equations are a set of coupled integro-differential equations which can be solved numerically for atoms but it is not possible to solve them for molecules because of the absence of spherical symmetry.

However, Roothaan (1951) showed that the method can be applied to molecules if the orbitals ψ_i are expressed as linear combinations of a set of functions $\{\chi_s\}$

$$\psi_i = \sum_s c_{is} \chi_s. \tag{5.9}$$

The expansion coefficients c_{is} are then obtained by an iterative method, to minimize W_{approx} as before. The set of functions $\{\chi_s\}$ is known as the basis set. This fits in with our more elementary concept of the LCAO (linear combination of atomic orbitals) approximation for molecular orbitals if the basis set consists of the usual core and valence atomic orbitals. For example, in the case of methane, the basis set would consist of the 1s, 2s and 2p orbitals on carbon and 1s orbitals on each of the hydrogen atoms. Such a basis set is known as a minimal basis set. At the other extreme, if we choose a sufficiently large and flexible set of functions $\{\chi_s\}$, then the orbitals ψ_i obtained will approximate to the solutions of the Hartree–Fock equations. For a basis set of m functions, the calculation will give $n/2$ occupied molecular orbitals and $m - n/2$ unoccupied or virtual orbitals in the case of a closed shell molecule containing n electrons. The method has been extended to open shells (Roothaan 1960) and we will assume below that for a given molecule the appropriate equations have been solved.

Early calculations used nodeless Slater functions

$$\chi_{nlm} = N_{nlm} r^{n-1} \exp(-\zeta r) Y_{lm}(\theta, \phi), \qquad (5.10)$$

where n, l and m are the usual atomic quantum numbers, N_{nlm} is a normalization factor and $Y_{lm}(\theta, \phi)$ is a spherical harmonic. The factor ζ in the exponent of the exponential function is known as the orbital exponent and there are rules for choosing suitable values. A minimal basis set will not give a very good wavefunction and a significant improvement can be obtained by using twice the number of basis functions in a double zeta basis set (so called because one has twice the number of orbital exponents ζ as in the case of a minimal basis set). A double zeta basis set for methane consists of two 1s functions, two 2s functions and two sets of 2p functions for carbon and two 1s functions for each hydrogen atom. Details of suitable orbital exponents are given by Schaefer (1972). However, such a basis set does not allow for the polarization of the electron cloud of one atom by a neighbouring atom during bond formation. In order to obtain an accurate description of the geometries and potential surfaces of molecules it is essential to include in the basis set additional functions to allow for this polarization effect. These functions are known as polarization functions and include p-functions on hydrogen atoms and d-functions on heavy atoms. Addition of these functions to a double zeta basis set gives what is known as a double zeta plus polarization set (DZP).

The calculation of the required molecular integrals is difficult and time consuming if exponential functions of the type given in equation 5.10 are used, and most molecular calculations are now done with basis sets of Gaussian functions:

$$\chi_{nlm} = N_{nlm} \exp(-\alpha r^2) Y_{lm}(\theta, \phi). \qquad (5.11)$$

The calculation of the required integrals is much faster with Gaussian functions but the penalty is that the functions are less capable of giving an accurate representation of the wavefunction. Thus, larger basis sets are required and a compromise has to be made between accuracy and computational efficiency in the iterative calculation of the molecular orbitals. This compromise is achieved by the use of contracted Gaussian basis sets in which each term in the expansion of equation 5.9 is written as a linear combination of Gaussian functions

$$\chi_s = \left(\sum_{i=1}^{n} d_{is} \exp(-\alpha_i r^2) \right) Y_{lm}(\theta, \phi). \qquad (5.12)$$

The coefficients d_{is} and orbital exponents α_i are fixed in any one calculation. This means that we can use a large number of Gaussian functions $\exp(-\alpha_i r^2)$ without increasing the actual number of basis functions used in the SCF calculation. Many methods have been proposed for the construction of contracted Gaussian basis sets for molecular calculations and we restrict ourselves to a brief discussion of a few of the more widely used basis sets. One approach is to expand a Slater function in terms of a linear combination of n Gaussian functions (Hehre *et al.* 1969). The expansion coefficients d_{is} and exponents α_i are chosen to give the best fit to the Slater orbital. The basis set resulting from the expansion of a Slater function in terms of n Gaussian functions is referred to as an STO–nG basis. Tabulations of the expansion coefficients and orbital exponents are given by Hehre *et al.* (1969) and by Stewart (1970). Although this expansion can be used for all of the orbitals in a DZ or DZP basis set, the use of STO–nG basis sets is usually restricted to calculations with minimal basis sets. For extended basis sets greater flexibility can be achieved by taking an uncontracted set of Gaussian functions obtained by optimizing the orbital exponents in an atomic self-consistent field calculation and then determining the contraction coefficients in a further atomic calculation. The determination and use of such basis sets have been reviewed by Dunning and Hay (1977) who give recommended contraction schemes. Other contraction schemes have been proposed by Pople and his coworkers. For example, a 6–31G basis for first-row atoms consists of an s-type inner shell function expressed as a combination of six Gaussian functions, a set of valence s- and p- type functions each expressed as a combination of three Gaussian functions and an outer set of s- and p- functions each consisting of a single Gaussian function (Binkley *et al.* 1980). A 6–31G basis augmented by d-polarization functions on the heavy atoms is designated 6–31G* whereas a 6–31G** basis has, in addition, p-functions on hydrogen atoms (Hariharan and Pople 1973).

Realistic results for potential surface calculations can be obtained only if the basis set is adequate for the purpose. A DZP basis is usually adequate for the calculation of surfaces for ground states and valence excited states. In molecules where mixing between valence and Rydberg orbitals may be important, it is essential that the basis set includes diffuse Rydberg functions. For states which yield a negative ion on dissociation it is necessary to include additional diffuse functions in order to obtain the correct asymptotic description.

As mentioned above, the molecular orbital wavefunction does not usually give a realistic description of potential surfaces. One of the obvious criteria to be satisfied in a potential surface calculation is that it gives the correct asymptotic dissociation limits. Also, the calculation should give a comparable description of the surface at all geometries. The molecular orbital wavefunction does not in general dissociate correctly. A well known example is that of the H_2 molecule which is predicted to dissociate into $H^+ + H^-$ rather than into two hydrogen atoms. This failure to dissociate correctly stems from the assumption that the wavefunction for a closed shell molecule can be written in terms of a set of doubly occupied molecular orbitals

$$\Phi = |\psi_1\alpha \; \psi_1\beta \; \psi_2\alpha \; \psi_2\beta \; \dots \; \psi_{n/2}\alpha \; \psi_{n/2}\beta|. \tag{5.13}$$

If dissociation is expected to lead to the electrons in orbital ψ_i ending up on different fragments A and B, then clearly a function of this form will be inadequate. In the limit of large R, ψ_i will be an orbital on either fragment A or fragment B and both electrons will end up on A or B respectively.

This difficulty can be resolved by partially relaxing the constraint of doubly occupied orbitals by rewriting the wavefunction for the pair of electrons $\psi_i\alpha \; \psi_i\beta$ in terms of two non-orthogonal orbitals ψ_{ia} and ψ_{ib} with singlet coupled electron spins

$$\Phi = |\psi_1\alpha \; \psi_1\beta \; \dots \; \psi_{ia}\psi_{ib}(\alpha\beta - \beta\alpha) \; \dots \; \psi_{n/2}\alpha \; \psi_{n/2}\beta|. \tag{5.14}$$

As R increases, ψ_{ia} transforms into an orbital on fragment A and ψ_{ib} into an orbital on fragment B and thus dissociation can be described correctly. This is the basis of the generalized valence bond method which is capable of giving potential surfaces which are qualitatively correct. For further details the reader should consult the review by Bobrowicz and Goddard (1977).

Even in cases where the wavefunction dissociates correctly the molecular orbital method does not give comparable accuracy at all points on the surface because it is based on the independent particle method which neglects electron correlation between electrons of opposite spin. The energy of the Hartree–Fock wavefunction differs from the exact non-relativistic energy by an amount known as the correlation energy. Since this will in general differ from one part of the surface to another, the calculated surface will not be parallel to the true surface at all points. In order to obtain realistic potential surfaces it is necessary to take into account electron correlation. In general this will at the same time ensure that the wavefunction dissociates correctly.

5.2.2. The Method of Configuration Interaction

The method of configuration interaction (CI) is perhaps the most generally applicable method for generating from a molecular orbital wavefunction a function which makes proper allowance for electron correlation. An outline of the method is given below but for a more detailed treatment the reader should consult Schaefer (1972) or Shavitt (1977). Let Φ_0 represent the SCF–MO wavefunction for an n-electron system in which $n/2$ orbitals are doubly filled. The SCF–MO calculation using a basis set of m functions gives a further $m - n/2$ virtual orbitals. By exciting electrons from the occupied orbitals to the virtual orbitals we can generate further determinantal functions. These will not all be of the same spin and symmetry as Φ_0 but where necessary linear combinations can be formed which have the required spin and symmetry. Let the set of all such configuration state functions (CSF) be represented by $\{\Phi_i\}$, $i = 1, \ldots, N$. The method of configuration interaction applies the variation method to a linear combination of the CSF's Φ_i, $i = 0, 1, \ldots, N$,

$$\Phi = \sum_{i=0}^{N} c_i \Phi_i. \qquad (5.15)$$

The lowest root from the secular determinant will give the best wavefunction that can be obtained for the ground state with the basis set used to calculate Φ_0. No other wavefunction using the same basis set will result in a lower energy. Higher roots are also upper bounds to the exact energies for excited states of the same spin and symmetry. Electron correlation is taken into account and the wavefunction dissociates correctly because even though Φ_0 may not correspond to the dissociation fragments, the CI wavefunction will include configurations which do correspond to those fragments. However, CI cannot correct for deficiencies in the basis set and accurate results will be obtained only if a realistic basis set is used.

The method outlined above is conceptually very straightforward but in most cases the number of configurations resulting from all possible excitations from Φ_0 is too large to handle and some limitation has to be imposed on the excitations considered. In limiting the number of configurations we have to remember that we require all parts of the surface to be described with comparable accuracy. If we are interested in several surfaces they should be treated on a comparable basis with particular attention being given to ensure correct asymptotic behaviour in all cases.

If two CSFs, Φ_i and Φ_j, differ by more than two orbital occupancies then it can be shown that the integral

$$\int \Phi_i H \Phi_j \, d\tau$$

is zero. Thus configurations which differ from Φ_0 by the excitation of more than two electrons cannot interact directly with Φ_0 and will be expected to have only a minor effect on the energy. An analysis by perturbation theory has shown that double excitations, in which two electrons are excited to virtual orbitals, play a dominant role. For a closed shell molecule, Brillouin's theorem states that for single excitations Φ_i, the integral

$$\int \Phi_0 H \Phi_i \, d\tau$$

is zero when Φ_0 is the SCF–MO wavefunction for the ground state. Thus for closed shell molecules, single excitations would not be expected to be very important. Although they have little effect on the energy, they can be very important in the calculation of molecular properties and should therefore be included. This is illustrated by the calculation of the dipole moment for CO. The SCF–MO wavefunction gave a dipole moment which was opposite in sign to the experimental value. CI studies showed that it was necessary to include single excitations in order to get a dipole moment having the correct sign (Grimaldi *et al.* 1967). Thus a reasonable initial approximation is to limit the configuration list to single and double excitations from Φ_0. However, in order to get a balanced treatment of several surfaces and to describe dissociation correctly one should also include single and double excitations with respect to those configurations which describe the excited states and those which are required to describe correctly the dissociation asymptotes. For accurate work the calculation should include such excitations from all configurations which have a coefficient c_i larger than 0.1 in the final CI wavefunction. This usually involves making several preliminary calculations in order to determine which configurations should be included in the multiple reference set of configurations $\{\Phi_k\}$ from which single and double excitations are generated.

In all but the smallest of molecules this will still result in a prohibitively large number of configurations. Some saving can be effected by keeping inner shell orbitals doubly occupied and by excluding excitations to the corresponding high energy virtual orbitals. This does not usually involve any loss of accuracy. The configuration lists can be reduced still further by imposing further limitations on the excitations considered. However, it is preferable not to restrict too drastically the classes of excitations included because important excitations may be excluded. An alternative approach is to generate all single and double excitations from a multiple set of reference configurations $\{\Phi_k\}$ and then truncate the configuration space by the elimination of the less important configurations. These configurations can be identified by calculating a zero-order CI wavefunction from the set of reference configurations $\{\Phi_k\}$ and then estimating by perturbation theory the energy lowering which would result from adding a particular configuration. All configurations arising from $\{\Phi_k\}$ are screened in this way and those which contribute less than a certain threshold energy, which is typically between $10^{-4} E_h$ and $10^{-5} E_h$ ($1 E_h = 4.359\,814 \times 10^{-18}$ J), are eliminated. The CI wavefunction is then calculated using the set of selected configurations. By performing the calculation at several thresholds it is possible to extrapolate the energy to zero threshold (Buenker and Peyerimhoff 1974). Objections may be made that this approach is rather unsatisfactory for potential energy surface calculations because a different configuration list is used for each point on the surface. However, the use of an extrapolation procedure should reduce any errors to a minimum.

During the past decade much effort has been devoted to the development of methods which are capable of handling very large configuration lists without having to resort to truncation and selection procedures. A discussion of these methods is beyond the scope of this chapter and the reader is referred to articles by Roos and Siegbahn (1977, 1980) for a full discussion. A complete CI calculation has been reported for the water molecule using a double-zeta basis set resulting in 256 473 configurations (Saxe *et al.* 1981).

It is generally necessary to use large configuration spaces in CI in order to obtain realistic results. This is partly because the virtual orbitals used for the generation of the excited configurations have not been subjected to any optimization but are merely a set of orbitals which are orthogonal to the occupied orbitals. The multiconfiguration SCF (MC–SCF) method attempts to obtain good results with more modest configuration lists by optimizing virtual orbitals as well as the occupied orbitals. The method uses a compact multiconfiguration wavefunction of the form

$$\Phi = \sum_i a_i \Phi_i, \tag{5.16}$$

where, in addition to optimizing the linear variational parameters a_i, the molecular orbital coefficients in the orbitals ϕ_i are simultaneously optimized. This double optimization is much more difficult to implement than the successive optimization involved in a SCF–MO calculation followed by a conventional CI calculation (see Olsen *et al.* 1983). However, if the configuration list is chosen carefully, potential surfaces are obtained which are parallel to those from more extensive CI calculations. In the optimized valence configuration scheme of Wahl and Das (1970, 1977), the assumption is made that in order to calculate realistic potential surfaces it is only necessary to include those configurations required to describe the changes in electronic structure which occur on molecule formation. These include those configurations which have to be added to the SCF–MO function to ensure correct dissociation and those which are important in the molecule but formally vanish at the dissociation limit.

5.2.3. *Valence Bond Theory*

An alternative approach to molecular electronic structure is that of valence bond theory. Molecular orbital theory considers the electronic wavefunction in terms of molecular orbitals which extend over the whole molecule and which have the same symmetry as the molecule. The identity of individual atoms is thus lost. By contrast, in valence bond theory the wavefunction is constructed from products of atomic wavefunctions. The use of valence bond theory for the construction of potential energy surfaces is thus appealing because correct dissociation is built into the model. We introduce the discussion of valence bond theory by a brief summary of the Heitler–London description of the hydrogen molecule. Fuller discussions can be found in Coulson (1961) and Slater (1963). If two hydrogen atoms with wavefunctions ψ_A and ψ_B are at infinite separation there is no interaction between them, and the total wavefunction can be written as the product of the individual wavefunctions

$$\psi = \psi_A \psi_B. \tag{5.17}$$

The Heitler–London method assumes that as the atoms are brought together, equation 5.17 is still a reasonable starting point for the construction of the molecular wavefunction. The indistinguishability of the electrons has to be taken into account and electron spin included. When this is done, it can be shown that the following form of wavefunction leads to a bound H_2 molecule

$$\psi = [\psi_A(1)\psi_B(2) + \psi_A(2)\psi_B(1)] \frac{1}{\sqrt{2}} [\alpha(1)\beta(2) - \alpha(2)\beta(1)], \tag{5.18}$$

where ψ_A and ψ_B are atomic orbitals on atoms A and B, respectively. Even a relatively crude function of this type can give a reasonably good description of the potential energy curve for H_2. Since the wavefunction is of the form appropriate to two separated atoms, the potential curve has the correct asymptotic behaviour.

This approach can be generalized to more complicated systems. However the valence bond method has not been widely used in the calculation of potential energy surfaces because of the computational difficulties arising from the non-orthogonality of atomic orbitals on different centres. There have been some recent developments in valence bond theory (see Gerratt 1974) which have improved the situation and a number of calculations have been reported in recent years.

One approach (Balint-Kurti and Karplus 1974) uses a set of approximate antisymmetrized many-electron atomic eigenfunctions Φ_i^A for all atomic states i on atom A to be included in the calculation. Products of these atomic functions, including at least one such function for each atom, are taken and, in order to satisfy the Pauli principle, antisymmetrized with respect to electrons which were originally assigned to different atoms. The resulting functions are known as composite functions Φ_r,

$$\Phi_r = \mathcal{A}[\Phi_i^A \Phi_j^B \Phi_k^C \ldots], \tag{5.19}$$

where $\mathcal{A}$ is a partial antisymmetrizer operator. The functions Φ_r may not have the correct spin properties so linear combinations Φ_r^{SM} are taken which have a well defined spin angular momentum quantum number S. The functions Φ_r^{SM} are then used in a variational calculation

$$\psi = \sum_r c_r \Phi_r^{SM}. \tag{5.20}$$

As well as having the advantage that the wavefunction will dissociate correctly, one can use chemical intuition to make a sensible choice of composite functions. It is possible with this method to obtain good general descriptions of ground and excited states with a much smaller set of functions Φ_r^{SM} than would be required in a comparable CI calculation. Several applications have been reported for diatomic molecules and for potential energy surfaces for simple chemical reactions (Balint-Kurti and Yardley 1977). In addition to using the method in a purely *ab initio* sense in which no approximations are made in the calculation, it is possible to use empirical atomic data to correct for known atomic errors resulting from the use of approximate atomic eigenfunctions Φ_i^A.

Valence bond theory is also important in the context of potential surface calculations because it forms the basis of several semi-empirical schemes to be discussed below.

5.2.4. *Some Examples of* Ab Initio *Calculations of Potential Surfaces*

In this section we discuss some examples from the recent literature of the application of *ab initio* methods to the calculation of potential energy surfaces. The first two examples are concerned with comparison between theory and spectroscopy whereas the remaining two examples are relevant to reaction dynamics.

5.2.4.1. LiH

As an example of the application of the CI method to a diatomic molecule, we consider the calculation of Partridge and Langhoff (1981) for the $X^1\Sigma^+$, $A^1\Sigma^+$ and $B^1\Pi$ states of LiH. This molecule has only four electrons and can therefore be treated to high accuracy. However, the $B^1\Pi$ state has a binding energy of only $0.035\,\text{eV}$ $(288\,\text{cm}^{-1})$ and the accurate calculation of the potential curve for this state is therefore a severe test. An extensive Slater basis set was used which resulted in a molecular orbital space of 22 σ orbitals, 12 π orbitals and 7 δ orbitals. Molecular orbitals for the $^1\Pi$ state were used in the CI calculations because they also gave a reasonable representation of the $^1\Sigma^+$ states. A set of 12 reference configurations was used for the $^1\Sigma^+$ calculation whereas the single configuration $1\sigma^2 2\sigma 1\pi$ was found to be adequate for the $^1\Pi$ state. These reference sets included all configurations having a coefficient of larger than 0.05 in the final CI wavefunction. The CI included all single and double excitations resulting in an energy lowering of more than $1\times10^{-7}\,E_h$. At such a low threshold the rejected configurations contribute virtually nothing to the total energy. The calculated curve for the $X^1\Sigma^+$ state is in excellent agreement with the 'experimental' curve obtained by joining the RKR potential obtained from the low vibrational levels to a long range potential including R^{-6}, R^{-8} and R^{-10}

terms. The calculated dissociation energy is $316\,\text{cm}^{-1}$ less than the experimental value of $20\,288\,\text{cm}^{-1}$. Vibrational energy levels were calculated by numerical integration of the radial Schrödinger equation using a potential curve obtained by fitting the *ab initio* energies. Comparison with experimental vibrational energy levels, which have been observed up to $v = 11$, indicates that discrepancies in the vibrational spacings are always less than $12\,\text{cm}^{-1}$. The calculations were less accurate for the $A\,^1\Sigma^+$ state because the basis set contained too few orbitals of π symmetry on lithium. Nevertheless, the calculated vibrational separations agree with experiment to $15\,\text{cm}^{-1}$ or better in all cases. The calculated dissociation energy was $360\,\text{cm}^{-1}$ larger than the experimental value of $8682\,\text{cm}^{-1}$. The calculated $\tilde{D}_e$ value for the $B\,^1\Pi$ state is $284\,\text{cm}^{-1}$ which is in very good agreement with experiment. Solution of the radial Schrödinger equation using the calculated potential function yields three vibrational levels in agreement with experiment and there is excellent numerical agreement for the vibrational spacings.

5.2.4.2. PH_2

It is also possible to obtain good agreement with spectroscopic data for triatomic molecules. Perić *et al.* (1979) have investigated the $\tilde{A}\,^2A_1 - \tilde{X}\,^2B_1$ transition in the PH_2 radical using an extensive Gaussian basis set consisting of 48 contracted Gaussian functions. One reference configuration for each state was found to be adequate for CI calculations of the angular potential curves (in which the bond length was fixed at the experimental equilibrium value for the ground state) and of stretching potential curves for relatively small distortions from the potential minimum. About 6000 configurations resulted from the generation of all single and double excitations and the use of a selection procedure reduced the configuration space to about 2000. The vibrational energy levels were calculated by a variational method using the theoretical potential surface. Discrepancies from experiment were less than $1°$ for bond angle, less than $0.001\,\text{Å}$ for bond lengths and less than $25\,\text{cm}^{-1}$ for bending vibration wavenumbers. The error in the calculated energy difference between the 000 levels of the two potential surfaces was $0.06\,\text{eV}$ $(480\,\text{cm}^{-1})$. The calculations were made within the framework of the Born–Oppenheimer separation and could not, therefore, account for perturbations in the vibrational levels of the upper state arising from the Renner–Teller effect. The $\tilde{A}$ and $\tilde{X}$ states become the two components of a $^2\Pi$ state for linear configurations. There are no direct experimental values for the stretching vibration wavenumbers which can be compared with the theoretical predictions. The intensity distribution was calculated for transitions between various vibrational levels of the bending mode and almost quantitative agreement was obtained with the observed intensities of the long progression.

5.2.4.3. H_3

The simplest reactive system, from a theoretical point of view, is the hydrogen exchange reaction

$$H + H_2 \rightarrow H_2 + H, \qquad (5.21)$$

which has received a lot of attention from the earliest days of quantum mechanics (Truhlar and Wyatt 1976, 1977). Very accurate calculations have been made for the collinear surface (Liu 1973) and for the three-dimensional surface (Siegbahn and Liu 1978). The earlier calculation used a 4s3p2d Slater basis on each hydrogen atom. This was shown to give good agreement with exact calculations for H_2. Full CI would have resulted in 14 949 configurations which was, at that time, prohibitively large. The configuration list was truncated and the full CI energies estimated by extrapolation. A total of 137 geometries were considered and the final surface was considered to be between $3.3 \, kJ \, mol^{-1}$ and $0.8 \, kJ \, mol^{-1}$ above the exact clamped nuclei surface ($1 \, kJ \, mol^{-1} \approx 84 \, cm^{-1}$). The errors result from shortcomings in the atomic basis set and not from limitations in the CI. Along the reaction path the deviation from the exact surface was estimated to be between 2 and $3 \, kJ \, mol^{-1}$. The calculations of Siegbahn and Liu (1978) employed a 4s3p1d Gaussian basis set but they considered that the use of this smaller basis set has very little effect on accuracy. A direct CI method was used with some truncation of the molecular orbital basis. In the region of the saddle point the resulting surface was thought to differ from the true surface by less than $0.4 \, kJ \, mol^{-1}$. On the basis of comparison of their calculations with earlier less extensive work they concluded that the calculation of barrier heights to an accuracy of less than $4 \, kJ \, mol^{-1}$ is likely to continue to be very costly but other properties of the saddle point region, reaction path and shape of the potential surface can be obtained accurately with more modest CI. A very accurate functional fit to the *ab initio* surface has been reported by Truhlar and Horowitz (1978).

5.2.4.4. The reaction $C^+ + H_2 \rightarrow CH^+ + H$

Theoretical potential energy surfaces play an invaluable role in the interpretation of the dynamics of reactions even though the surfaces may not be accurate enough or sufficiently extensive for dynamical calculations. In this section we discuss the interpretation of some aspects of the reaction of $C^+(^2P)$ with H_2. At low relative energies it appears that the reaction

$$C^+(^2P) + H_2 \rightarrow CH^+(X \, ^1\Sigma^+) + H \qquad (5.22)$$

proceeds through a collision complex whereas at energies above $5 \, eV$ the mechanism is direct. On the basis of high-resolution velocity analysis of the product CH^+, it has been suggested that two additional reactions are important

$$C^+(^2P) + H_2 \rightarrow CH^+(a\,^3\Pi) + H \qquad (5.23)$$

$$C^+(^2P) + H_2 \rightarrow CH^+(A\,^1\Pi) + H. \qquad (5.24)$$

Chemiluminescence has been observed from the $CH^+(A\,^1\Pi)$ produced in the reaction as well as emission from $CH(A\,^2\Delta)$ and $CH^+(B\,^1\Delta)$ produced in the reactions

$$C^+(^2P) + H_2 \rightarrow CH(A\,^2\Delta) + H^+ \qquad (5.25)$$

and

$$C^+(^2P) + H_2 \rightarrow CH^+(B\,^1\Delta) + H. \qquad (5.26)$$

Reaction 5.24 has been particularly thoroughly studied but its mechanism could not be properly understood in the absence of potential energy surfaces. Sakai *et al.* (1981), using a Gaussian basis which included polarization functions, have calculated CI potential energy surfaces for this system with the prime intent of elucidating the mechanism of reaction 5.24. They included all the 'valence' configurations arising from five valence electrons in the six valence orbitals. They considered collinear, perpendicular (C_{2v}) and C_s geometries. The C_s geometries are defined as in figure 5.1.

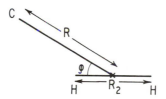

Figure 5.1. Coordinates used for reaction of C^+ with H_2.

Correlation diagrams derived according to the considerations of section 3.2 are shown in figures 5.2 and 5.3 for the collinear case and for the B_2 and A_1 states in C_{2v} symmetry.

In the collinear case, states of CH_2^+ of $^2\Pi$ and $^2\Sigma^+$ symmetry correlate with the reactants. The $^2\Sigma^+$ state correlates adiabatically with $CH^+(^1\Sigma^+)$ and the calculated potential surface indicates that reaction 5.22 can occur by a direct mechanism on this surface. The $^2\Pi$ state of CH_2^+ correlates diabatically with $CH(^2\Pi) + H^+$ but an avoided crossing with a second $^2\Pi$ state (correlating with $C(^3P) + H_2^+$) leads adiabatically to the products $CH^+(^3\Pi) + H$. Reaction 5.23 can be interpreted in terms of motion on this surface. There is a second avoided crossing with the $3\,^2\Pi$ state of CH_2^+ (correlating with $C(^1D) + H_2^+$) so that the $2\,^2\Pi$ state correlates adiabatically

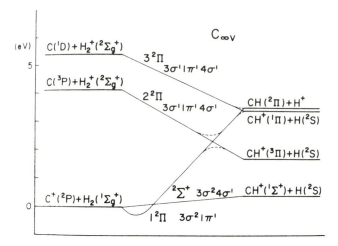

Figure 5.2. Correlation diagram for collinear geometries for reaction of C^+ with H_2. (Reproduced from Sakai *et al.* 1981.)

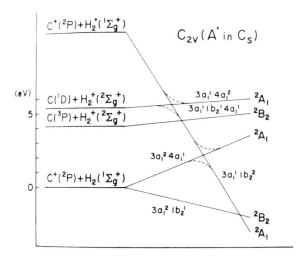

Figure 5.3. Correlation diagram for $^2A_1$ and $^2B_2$ states for reaction of C^+ with H_2. (Reproduced from Sakai *et al.* 1981.)

with $CH^+(^1\Pi) + H$. However, the separation between the calculated collinear $1\,^2\Pi$ and $2\,^2\Pi$ surfaces of CH_2^+ is too large for it to be reasonable to interpret reaction 5.24 in terms of a non-adiabatic transition between them. In C_{2v} symmetry it is the $2\,^2A_1$ surface which correlates with the products $CH^+(A\,^1\Pi) + H$ when a hydrogen atom is pulled away. The calculated $1\,^2A_1$

potential surface (figure 5.4) is initially strongly repulsive with a saddle point on a ridge arising from the avoided intersection with the $2\,^2A_1$ surface. Beyond the saddle point the surface descends rapidly into a deep potential well corresponding to the equilibrium geometry of the ground state CH_2^+.

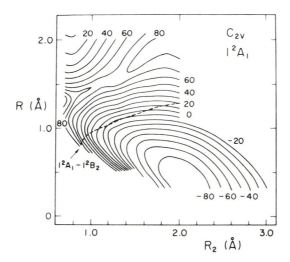

Figure 5.4. Contour diagram for $1\,^2A_1$ surface for reaction of C^+ with H_2. (Reproduced from Sakai *et al.* 1981.)

The $2\,^2A_1$ surface (figure 5.5) is initially attractive and has a narrow groove at geometries corresponding to the ridge in the $1\,^2A_1$ surface. The separation between the two surfaces is quite small (about $33\,kJ\,mol^{-1}$) and it is thought that non-adiabatic transitions between these surfaces may play an important role in reaction 5.24. As the symmetry relaxes from C_{2v} to C_s, the two surfaces (now $2\,^2A'$ and $3\,^2A'$) remain fairly close in energy for angles of ϕ (see figure 5.1) from 90° to 60°. For smaller values the energy separation increases. Non-adiabatic transitions can occur for motion perpendicular to the seam of the avoided intersection for angles in the range of ϕ from 90° to 60°. When the system has reached the upper $3\,^2A'$ surface it will have used up virtually all the endothermicity of the reaction. It will probably hit a repulsive wall on the $3\,^2A'$ surface and run downhill to form products $CH^+(^1\Pi)+H$. The conditions for this non-adiabatic process to occur are quite restrictive and it is therefore reasonable that reaction 5.24 is a minor process with a small cross-section.

Potential wells occur on the $1\,^2B_2$, $1\,^2A_1$ and $1\,^2B_1$ surfaces and motion on these surfaces may be responsible for the observation of a complex mechanism at low energies.

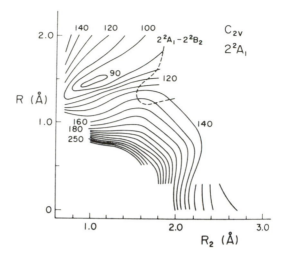

Figure 5.5. Contour diagram for $2\,^2A_1$ surface for reaction of C^+ with H_2. (Reproduced from Sakai *et al.* 1981.)

5.3. Semi-Theoretical Methods: Use of Effective Core Potentials

The *ab initio* methods discussed in the previous section explicitly include all the electrons in the molecule. The size of the calculation increases rapidly as the number of basis functions is increased. For example, the number of two-electron integrals which have to be generated and manipulated is roughly proportional to the fourth power of the size of the basis set. Thus it is only possible to make good quality *ab initio* CI calculations, in which an adequate basis set is used, for relatively small molecules containing light atoms. However, a chemist divides the electrons of a molecule into two groups—the core electrons from completely filled inner shells and the valence electrons. It is the valence electrons which are primarily responsible for chemical bonding and electrons from inner shells play virtually no part in the bonding. Intuitively one feels that it ought to be possible to devise methods which just consider the valence electrons and thus simplify calculations for heavier atoms. If this can be done, then a calculation for the I_2 molecule should be no more difficult or time consuming than a calculation for F_2.

It is not possible to do this without making some approximations but it is possible to reduce the computational effort in a CI calculation in which the core orbitals are frozen. We write the total electronic wavefunction

$$\psi = \mathcal{A}(\psi_{\text{core}}\psi_{\text{valence}}), \qquad (5.27)$$

where $\mathcal{A}$ is the antisymmetrizer and ψ_{core} and ψ_{valence} are wavefunctions for the core and valence electrons respectively. If ψ_{valence} is orthogonal to ψ_{core}, the total energy can be written as

$$E_{\text{total}} = E_{\text{core}} + \int \psi_{\text{valence}} H_{\text{v}} \psi_{\text{valence}} \, d\tau. \tag{5.28}$$

H_{v} is the valence Hamiltonian which is of the form

$$H_{\text{v}} = \sum_v \left(h_v + \sum_c g_{vc} \right) + \sum_{v>v'} \frac{1}{r_{vv'}}, \tag{5.29}$$

where h_v represents the one-electron energy of the valence electron v, g_{vc} is an operator representing the effect of core electron c on valence electron v and the term $1/r_{vv'}$ represents the interelectronic repulsion between valence electrons v and v'. Evaluation of the terms g_{vc} requires the two-electron integrals with respect to both core and valence electrons and so no saving is achieved at the SCF level. However significant savings in computational effort can be made in CI calculations in which inner shell orbitals are kept doubly occupied.

If we use an approximate valence Hamiltonian of the form

$$H_{\text{v}} \approx \sum_v [h_v + U^{\text{core}}(v)] + \sum_{v>v'} \frac{1}{r_{vv'}}, \tag{5.30}$$

where $U^{\text{core}}(v)$ is an effective core potential representing the effect of the core electrons on the valence electrons, then we can perform calculations which just consider the valence electrons. The simplest approach is to represent the effective core potential by a suitable functional form, or model potential, which may simply be a radial function or may be angle dependent. The adjustable parameters in the model potential may be chosen to fit all-electron calculations or alternatively reproduce experimental data. Details of the most frequently used model potentials are given by Dixon and Robertson (1978) and by Kahn et al. (1976).

A more general and flexible approach is to determine the effective core potential $U^{\text{core}}(v)$ from atomic all-electron calculations. This method avoids the necessity of constraining $U^{\text{core}}(v)$ to a particular functional form. One approach of this type has been described in detail by Kahn et al. (1976) who have used it in the calculation of potential curves for hydrides of halogens and for diatomic halogen compounds. However some discrepancies relative to all-electron calculations were noted and an alternative approach has been proposed (Christiansen et al. 1979). Relativistic effects are very important for heavy atoms and should be taken into account in the construction of effective core potentials for heavy atoms. Methods have been proposed by Kahn et al. (1978) and by Lee et al. (1977) in which relativistic atomic wavefunctions have been used in the calculation of effective core potentials and numerous applications to the calculation of potential curves have been reported. Comparison with all electron calculations shows that comparable results can be obtained in calculations with effective core potentials and we can expect to see these methods applied very widely in the future.

5.4. Semi-Empirical Methods

Although it is now possible to calculate by *ab initio* methods accurate potential curves for diatomic molecules and portions of the potential surface for small polyatomic systems, such methods are too expensive to be used routinely for many systems of chemical interest. There are two problems associated with the application of these methods to the calculation of potential surfaces. The time taken for the calculation of each point on the surface increases with the size of the molecule and also the number of geometrical variables increases as $3N-6$, where N is the number of atoms. Thus much effort has been devoted to devising semi-empirical methods which can be applied to complex molecules or which can be used to model potential surfaces for reactive systems. Many semi-empirical schemes based on the molecular orbital method have been proposed for valence electrons. Although some of these, e.g. MINDO/3 (Dewar 1977), are of considerable value in the calculation of potential surfaces for organic reactions, they are not generally well suited for the representation of potential surfaces for small molecules and for simple reactive systems and will not be considered further here. Two widely used methods which are based on the valence bond method are the London–Eyring–Polanyi–Sato function (LEPS) and Diatomics-in-Molecules (DIM) which will be discussed in some detail below. Other semi-empirical schemes have been proposed for particular problems such as the reaction of $F+H_2$ (Blais and Truhlar 1973) and the reactions of alkali atoms with hydrogen halides and diatomic halogens (Zeiri and Shapiro 1978).

5.4.1. The London–Eyring–Polanyi–Sato Function

A whole range of semi-empirical and empirical potential surfaces have been derived from a valence bond approach to the potential energy surface for H_3 which is relevant to the simplest of chemical reactions

$$H+H_2 \rightarrow H_2+H. \tag{5.31}$$

In equation 5.17 we wrote down the Heitler–London wavefunction for the H_2 molecule. The energy corresponding to this singlet wavefunction is

$$E_1 = \frac{Q_{ab} + J_{ab}}{1 + S_{ab}}, \tag{5.32}$$

where Q_{ab}, the Coulomb integral, and J_{ab}, the exchange integral, are given by

$$Q_{ab} = \int \phi_A(1)\phi_B(2) H \phi_A(1)\phi_B(2) \, d\tau \tag{5.33}$$

and

$$J_{ab} = \int \phi_A(1)\phi_B(2) H \phi_A(2)\phi_B(1) \, d\tau \tag{5.34}$$

and S_{ab} is the overlap integral

$$S_{ab} = \int \phi_A(1)\phi_B(1) \, d\tau. \tag{5.35}$$

The energy expression for the corresponding repulsive triplet state is

$$E_3 = \frac{Q_{ab} - J_{ab}}{1 - S_{ab}}. \tag{5.36}$$

Extending the method to H_3 gives a rather complicated energy expression which involves overlap integrals, a Coulomb integral, two-centre exchange integrals and a three-centre exchange integral. Since the derivation is based on a minimal basis set description of H_3, use of *ab initio* values for the integrals would not be expected to yield very good results. Potential surfaces for H_3 have been derived by this method using a combination of empirical data, some approximations and the explicit evaluation of overlap integrals (Porter and Karplus 1964, Pedersen and Porter 1967). Two-centre Coulomb and exchange integrals Q_{ab}, J_{ab}, etc., can be obtained by using the empirical potential energy functions for the singlet ground state E_1 and for the repulsive triplet state E_3 in equations 5.32 and 5.36 and solving the simultaneous equations for Q_{ab} and J_{ab}. Although the method gives results which are in agreement with good *ab initio* calculations, the accuracy of such methods is limited.

Neglect of overlap integral and the three-centre exchange integral leads to the well known London formula

$$E = Q_{ab} + Q_{bc} + Q_{ac} - [\tfrac{1}{2}(J_{ab} - J_{bc})^2 + \tfrac{1}{2}(J_{ab} - J_{ac})^2 + \tfrac{1}{2}(J_{bc} - J_{ac})^2]^{\frac{1}{2}}. \tag{5.37}$$

Early evaluation of this potential function assumed, erroneously, that the ratio Q_{ij}/J_{ij} was equal to 0.25 and was independent of internuclear distance. This led to the prediction of a small well at the top of the activation barrier. A more satisfactory approach is to obtain the Coulomb and exchange integrals empirically from the singlet and triplet curves for H_2. With neglect of overlap we now have from equations 5.32 and 5.36

$$E_1 = Q_{ab} + J_{ab}, \qquad E_3 = Q_{ab} - J_{ab}. \qquad (5.38)$$

Use of the resulting values of Q and J leads to a realistic potential surface for H_3 (Cashion and Herschbach 1964). This approach can be extended to other three-electron systems such as triatomic alkali metal surfaces.

These methods are only strictly applicable to systems involving three electrons in s orbitals because they are derived from the application of the valence bond method to H_3. However, the London formula does provide a function which has the right sort of characteristics for the potential energy surface for a reactive system. The function is capable of describing dissociation and the presence of an activation barrier correctly. It has been used as the starting point for the formulation of potential energy surfaces for reactive systems other than those for which it is strictly applicable. For the sake of continuity these developments will be discussed here although such surfaces really are in the category of empirical surfaces. In order to make some allowance for overlap, Sato modified the London formula (equation 5.37) by multiplication by a factor of $1/(1+k)$ where k is an adjustable parameter. The resulting function is usually referred to as the London–Eyring–Polanyi–Sato (LEPS) function. The Coulomb and exchange integrals are now related to the singlet and triplet diatomic curves by the relationships

$$E_1 = \frac{Q_{ab} + J_{ab}}{1 + k}, \qquad E_3 = \frac{Q_{ab} - J_{ab}}{1 - k}. \qquad (5.39)$$

The singlet potential curve is usually represented by a Morse function

$$E_1 = D_e[\exp\{-2\beta(R - R_e)\} - 2\exp\{-\beta(R - R_e)\}] \qquad (5.40)$$

and the triplet curve by an anti-Morse function

$$E_3 = \tfrac{1}{2}D_e[\exp\{-2\beta(R - R_e)\} + 2\exp\{-\beta(R - R_e)\}]. \qquad (5.41)$$

The parameter k can be chosen so that the resulting potential surface has a barrier height with a specific value.

A more flexible function, the extended LEPS function, has been proposed by Kuntz *et al.* (1966) in which three adjustable parameters k_{ab}, k_{bc} and k_{ac} are introduced to give the energy formula

$$E = \frac{Q_{ab}}{1 + k_{ab}} + \frac{Q_{bc}}{1 + k_{bc}} + \frac{Q_{ac}}{1 + k_{ac}} - \left[\frac{J_{ab}^2}{(1 + k_{ab})^2} + \frac{J_{bc}^2}{(1 + k_{bc})^2} + \frac{J_{ac}^2}{(1 + k_{ac})^2} \right.$$

$$\left. - \frac{J_{ab}J_{bc}}{(1 + k_{ab})(1 + k_{bc})} - \frac{J_{ab}J_{ac}}{(1 + k_{ab})(1 + k_{ac})} - \frac{J_{bc}J_{ac}}{(1 + k_{bc})(1 + k_{ac})} \right]^{\frac{1}{2}}. \qquad (5.42)$$

The LEPS and extended LEPS functions have been very widely used in classical trajectory calculations for reactive systems (see chapter 6). Muckerman (1981), for example, has reviewed the potential surfaces that have been used for the system $F + H_2$. In addition to being used to simulate the dynamics of actual reactions, the flexible form of the extended LEPS function makes it well suited to a theoretical study of the relationship between reaction dynamics and the topography of the potential surface. However, despite its usefulness, the extended LEPS function does not seem to have enough flexibility to give a good description of all parts of a polyatomic surface. Indeed, spurious hollows may occur in a surface forced to have approximately the desired barrier height and saddle point location.

5.4.2. *Diatomics-in-Molecules*

As mentioned above, the London formula is only strictly applicable to systems of three electrons in s orbitals. However, the use of empirical diatomic data in the construction of polyatomic potential energy surfaces is very appealing since it ensures a correct description of diatomic fragments. The method of diatomics-in-molecules is based on a more general valence bond approach to polyatomic molecules in which the energy is ultimately expressed in terms of the energies of atomic and diatomic fragments.

The n-electron Hamiltonian for a polyatomic molecule can be partitioned exactly into terms which are Hamiltonian operators for diatomic and atomic fragments (Ellison 1963)

$$H = \sum_{K}^{N} \sum_{L>K}^{N} H^{(KL)} - (N-2) \sum_{K}^{N} H^{(K)}, \qquad (5.43)$$

where $H^{(KL)}$ and $H^{(K)}$ are the Hamiltonian operators for the diatomic molecule KL and the atom K respectively. The wavefunction is a valence bond function built up from functions which are products of wavefunctions on the isolated atoms. Let $\xi_m^{(A)}(1,\ldots,n_A)$ represent a many electron wavefunction for n_A electrons on atom A. A product $\phi_m(1,\ldots,n)$ is taken of such functions including one for each atom A, B, ..., N.

$$\phi_m(1,\ldots,n) = \xi_m^{(A)}(1,\ldots,n_A)\xi_m^{(B)}(n_A+1,\ldots,n_A+n_B)$$
$$\ldots\xi_m^{(N)}(n-n_N+1,\ldots,n). \qquad (5.44)$$

Application of the antisymmetrizer $\mathscr{A}$ to ϕ_m gives a polyatomic basis function $\Phi_m(1,\ldots,n)$

$$\Phi_m(1,\ldots,n) = \mathscr{A}\phi_m(1,\ldots,n). \qquad (5.45)$$

The total wavefunction for the molecule is then written as a linear combination of the polyatomic basis functions

$$\psi(1,\ldots,n) = \sum_{m} c_m \Phi_m(1,\ldots,n) \qquad (5.46)$$

and the expansion coefficients are obtained by solution of the variational problem. The secular problem can be simplified substantially if linear combinations of Φ_m are taken which are spin eigenfunctions. The integrals required for the solution of the variational problem in equation 5.46 can be expressed in terms of the Hamiltonians for atomic and diatomic fragments

$$\int \Phi_i H^{(K)} \Phi_j \, d\tau \quad \text{and} \quad \int \Phi_i H^{(KL)} \Phi_j \, d\tau.$$

These integrals can in turn be expressed in terms of the energies of the atomic and diatomic fragments. This treatment is exact if a complete set of polyatomic basis functions $\{\Phi_m\}$ is used in equation 5.46. This is, of course, not possible and it is necessary to use a truncated set. At this point the assumption is made that even with a truncated set of polyatomic basis functions one can still express the required integrals in terms of atomic and diatomic energies. The way in which this is done is rather complicated and details will not be given here. Full details of the implementation of the method can be found in review articles by Kuntz (1979) and Tully (1977, 1980). In the case of H_3 use of a set of polyatomic basis functions built up from covalent structures involving H atoms in the ^2S state, the DIM method results in the London equation if overlap is neglected.

There are a number of reasons why the DIM method is appealing for the calculation of potential surfaces. The use of atomic and diatomic data in the calculation of the integrals with respect to the Hamiltonian ensures that the potential surface will dissociate correctly. However, in order to obtain realistic results it is important to choose a basis set of atomic functions $\{\xi_m^{(I)}\}$ that includes all the atomic states which are required to describe adequately the atom in any chemically bonded situation in the molecule being considered. A good description of the diatomic fragments can be ensured by using the best available experimental or *ab initio* data. Excited state potential surfaces can be calculated readily provided that a sufficiently large number of atomic states are included in the basis set. It is also relatively straightforward to include spin–orbit effects in the formalism. The method can be shown to give a satisfactory description of the energetics of bonding, valence and the directionality of chemical bonding. A number of case studies of the application of DIM potential surfaces to reaction dynamics are discussed by Tully (1980). Although the DIM method works well in many cases, there are some systems for which it seems to be incapable of giving potential surfaces of quantitative accuracy and thus the method must be applied with caution.

5.5. Empirical Potential Functions

Many of the methods discussed in the previous section yield potential energy surfaces in the form of a table of energies for given geometries. In any dynamical calculation it is necessary to evaluate the potential energy for any arbitrary geometry. Thus in order to use the calculated potential surface it is either necessary to interpolate between the calculated points or to fit them to some analytical function. Suitable analytical functions are not usually derived from any quantum mechanical theory of the bonding in the molecule but are chosen to give the best fit over the widest range of geometries. The extended LEPS function is, in general, not flexible enough to give a surface of sufficient accuracy for dynamical calculations. In addition to simply fitting a set of calculated points, there is also considerable effort being made to devise realistic analytical functions which can use a combination of *ab initio* data and spectroscopic data. There are also a number of empirical potential surfaces for specific systems in which a particular model has been chosen to suit the system being considered and as an example of this approach we discuss below the Rittner potential for ionic molecules.

The most general approach to interpolation between a set of calculated points is the use of spline fitting. This does not involve the choice of an analytical function which may or may not be uniformly satisfactory. Spline fitting requires continuity of the energy and of its first two derivatives, and hence ensures smoothness of the interpolated surface. The use of spline fitting in dynamical calculations has been discussed by McLaughlin and Thompson (1973) and by Sathyamurthy and Raff (1975), and further aspects of spline interpolation have been considered by Gray and Wright (1978). The method is capable of fitting satisfactorily a set of points but does have a number of disadvantages. For spline interpolation in three dimensions it is necessary to have a complete rectilinear grid of points. It is therefore necessary to have a coordinate system which represents the entire three-dimensional surface in a rectilinear form. The generation of the surface will probably require the calculation of a large number of points well away from any reaction coordinate. In a comparison of dynamical calculations using an analytical function and spline fits to it, Sathyamurthy and Raff (1975) found that although they did not get a point by point match in individual trajectories, there was good agreement for total cross-sections, energy partitioning and spatial distribution. However, a subsequent study of the dynamics of the $He + H_2^+$ reaction in which an *ab initio* surface was fitted by spline functions and by a diatomics-in-molecules calculation showed that the dynamics are very sensitive to small differences in the potential function (Sathyamurthy *et al.* 1977).

Several analytical functions have been proposed for the representation of potential surfaces. For collinear geometries for the A-B-C system, several workers have considered the surface to be generated by the rotation of a Morse function about a point (R_{AB}, R_{BC}) in the high plateau region corresponding to separated atoms. The function is of the form

$$V(r, \phi) = D(\phi)[(1 - \exp\{-\beta(\phi)[r_0(\phi) - r]\})^2 - 1]. \qquad (5.47)$$

The angle ϕ varies from $0°$ to $90°$ and the Morse parameters D, β and r_0 are functions of ϕ (see figure 5.6).

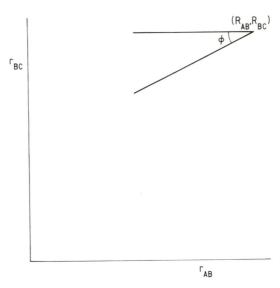

Figure 5.6. Definition of geometrical parameters for a rotated Morse function.

Thus for $\phi = 0°$ and $\phi = 90°$, we obtain the Morse functions for the diatomic fragments AB and BC. In the original formulation (Wall and Porter 1962) functional forms were given for D, β and $r - r_0$, but subsequent workers have chosen the values to fit the actual surface for a chosen value of ϕ (Bowman and Kuppermann 1975, Connor *et al.* 1975). Allowing D, β and r_0 to vary freely makes the function much more flexible and a better fit can be achieved.

Another analytical function which is based on the Morse potential is the switching function of Blais and Bunker (1962). For the reaction $A + BC \rightarrow AB + C$ they used the function

$$V = D_{AB}[1 - \exp\{-\beta_{AB}(R_{AB} - R_{AB}^0)\}]^2 + D_{BC}[1 - \exp\{-\beta_{BC}(R_{BC} - R_{BC}^0)\}]^2$$
$$+ D_{BC}[1 - \tanh(aR_{AB} + c)]\exp[-\beta_{BC}(R_{BC} - R_{BC}^0)]$$
$$+ D_{AC}\exp[-\beta_{AC}(R_{AC} - R_{AC}^0)], \qquad (5.48)$$

where R_{AB}^0 and R_{BC}^0 are the equilibrium bond lengths in AB and BC. The first two terms are simply the Morse potentials for AB and BC. The third term has the effect of reducing the attraction between B and C as A approaches and the fourth term represents the repulsive interaction between atoms A and C.

A more flexible function is the hyperbolic map function proposed by Bunker (see Bunker and Parr 1970). If $R_{AB} - R_{AB}^0 = x$ and $R_{BC} - R_{BC}^0 = y$, then the minimum energy path from reactants to products can be represented by a rectangular hyperbola

$$xy = u_0. \tag{5.49}$$

We can define a set of conjugate hyperbolae which intersect the reaction path at right angles by

$$v = \tfrac{1}{2}(y^2 - x^2). \tag{5.50}$$

The point of intersection of the two hyperbolae will depend on the value of v. If v is very negative, then the two hyperbolae will intersect in the entrance channel where R_{AB} is very large. Conversely, the intersection will occur in the exit channel (R_{BC} very large) if v has a large positive value. Thus the value of v gives a measure of the distance travelled along the reaction path. A parameter S is defined as the distance between the point (x, y) and the point (x_0, y_0) at which the conjugate hyperbola passing through (x, y) intersects the rectangular hyperbola. The potential function is a generalized Morse potential

$$V = F(\alpha, v) D(v)[\exp\{-2\beta(v)S\} - 2\exp\{-\beta(v)S\}], \tag{5.51}$$

where α is the angle between the AB and BC bonds. The angular function $F(\alpha, v)$ approaches unity as v becomes very negative or very positive.

A rather crude model for the changes in bonding in the collinear $A + BC \rightarrow AB + C$ reaction is the basis of the bond-energy bond-order (BEBO) method of Johnston (1966). The model assumes the transfer of an atom between two singly valent doublet atoms or radicals. The potential energy is written as the dissociation energy of the BC bond minus the sum of the energies of the partially formed AB bond and of the partially broken BC bond plus the energy of repulsion between atoms A and C. The energies of the partial bonds are expressed in terms of the bond dissociation energies D_{XY} and bond orders n_{XY}. Thus

$$V_{\parallel} = D_{BC} - D_{BC}(n_{BC})^{p_{BC}} - D_{AB}(n_{AB})^{p_{AB}} + V(R_{AC}), \tag{5.52}$$

where p_{XY} is a parameter. Since the AB bond is being formed while the BC bond is being broken we have the relation

$$n_{AB} + n_{BC} = 1. \tag{5.53}$$

The bond order is related to bond length by Pauling's relation

$$R_{XY} = R_{XY}^0 - C \ln n_{XY}, \tag{5.54}$$

where R_{XY}^0 is the equilibrium bond length and C is given the value of 0.26 Å. The repulsive interaction can be represented by an anti-Morse function. Equation 5.52 specifies the variation in energy along the reaction path and it is necessary to add to this a term $V_\perp$ which gives the variation in energy for motion perpendicular to the reaction path. In the harmonic approximation this is given by

$$V_\perp = \tfrac{1}{2}F_{AB}(R_{AB} - R_{AB}^p)^2 + \tfrac{1}{2}F_{BC}(R_{BC} - R_{BC}^p)^2, \qquad (5.55)$$

where R_{AB}^p and R_{BC}^p are the coordinates of the corresponding point on the reaction path and F_{AB} and F_{BC} are force constants. The method has been refined and extended by Garrett and Truhlar (1979) and by Schatz *et al.* (1981). Distortion perpendicular to the reaction path is described by a Morse potential and the mapping of the full surface involves the rotation of a Morse function about a swing point as in the rotated Morse potential method. The parameters can be chosen using the BEBO prescription to give an empirical potential or they can be chosen to fit a surface obtained by some other means. The latter approach is discussed by Wagner *et al.* (1981) in a detailed comparison of the use of the BEBO model with rotated Morse functions and the extended LEPS function.

These functions, and variants of them, have been used to represent potential energy surfaces for simple reactive systems in dynamical calculations. A more general problem is that of finding a suitable function which is capable of describing both the equilibrium configuration of a stable molecule and potential surfaces for reactive systems. Murrell and his coworkers (see Murrell *et. al.* 1984) have developed a relatively simple analytical function which can give a good description of potential surfaces for several different types of triatomic systems. The function can be used to fit *ab initio* data or alternatively the parameters can be chosen to fit the spectroscopic potential. The potential is expressed as a sum of two-body terms and a three-body term expressed as functions of the internuclear distances R_{AB}, R_{BC} and R_{AC}:

$$V(R_{AB}, R_{BC}, R_{AC}) = V_{AB}(R_{AB}) + V_{BC}(R_{BC})$$
$$+ V_{AC}(R_{AC}) + V_{ABC}(R_{AB}, R_{BC}, R_{AC}). \quad (5.56)$$

The three body-term is chosen so that it becomes zero when any one atom is pulled away to infinity and the two-body terms are simply diatomic potential functions. Thus the potential function has the correct asymptotic behaviour. The three-body potential is expressed in terms of displacements s_i from a suitably chosen configuration as

$$V_{ABC}(R_{AB}, R_{BC}, R_{AC}) = AP(s_1, s_2, s_3) \prod_{i=1}^{3} (1 - \tanh \tfrac{1}{2}\gamma_i s_i), \quad (5.57)$$

where A is a constant, P is a polynomial containing up to quartic terms and γ_i are adjustable parameters. The coefficients in the polynomial P are also regarded as variable parameters. The number of terms retained in P depends on the amount of data available. When the function is being fitted to spectroscopic data, the coefficients in the polynomial P are related to the force constants.

Comparison of calculated and experimental vibration frequencies has shown that a potential function of this type is capable of giving a good representation of the quartic force field in H_2O. The method has also been shown to account for multiple potential minima and to predict metastable minima correctly and has been extended to describe conical intersections between two surfaces of the same symmetry. For example, in the case of the $\tilde{X}\,^1A'$ and $\tilde{B}\,^1A'$ surfaces of H_2O, the adiabatic energies are obtained by calculating the eigenvalues of a 2×2 matrix. Each of the diagonal elements is given by a function of the form of equation 5.56 and the off-diagonal term is of the form

$$V_{12} = C \sin \text{HOH}, \tag{5.58}$$

where C is a three-body term.

The method can be generalized to tetra-atomic molecules by writing the potential as a sum of two-body and three-body terms, plus a four-body term. The four-body term is often relatively small and quite a reasonable description of the potential surface can be obtained by ignoring it.

A very successful model (Rittner 1951) for the potential energy function for an ionic molecule can be constructed by representing the potential as the sum of the classical electrostatic charge–charge, charge–induced dipole and induced dipole–induced dipole potentials with a repulsive Born–Meyer term for short bond lengths and an r^{-6} term to represent the long range van der Waals interaction. Thus for a diatomic molecule we have

$$V = -\frac{e^2}{r} - e^2 \frac{\alpha_1 + \alpha_2}{2r^4} - \frac{2e^2 \alpha_1 \alpha_2}{r^7} + A \exp\left(-\frac{r}{\rho}\right) + \frac{C}{r^6}, \tag{5.59}$$

where α_1 and α_2 are the polarizabilities of the two ions and A, ρ and C are constants. The exponential term represents the short-range repulsive interaction. The model can be generalized for polyatomic species by using classical electrostatics to calculate the induced dipole moment at one ion resulting from the electric field of the other ions (Lin *et al.* 1973).

Suggestions for Further Reading

P. W. Atkins (1983) *Molecular Quantum Mechanics*, 2nd edition, Oxford University Press, Oxford.

R. McWeeny and B. T. Sutcliffe (1969) *Methods of Molecular Quantum Mechanics*, Academic Press, New York.

W. G. Richards and D. L. Cooper (1983) *Ab Initio Molecular Orbital Calculations for Chemists*, Clarendon Press, Oxford.

H. F. Schaefer (1972) *The Electronic Structure of Atoms and Molecules: A Survey of Rigorous Quantum Mechanical Results*, Addison–Wesley, Reading MA.

H. F. Schaefer (1977) *Methods of Electronic Structure Theory*, Plenum Press, New York.

H. F. Schaefer (1984) *Quantum Chemistry: The development of ab initio methods in molecular electronic structure theory*, Oxford University Press, Oxford.

References

R. F. W. Bader and R. A. Gangi (1975) in *Theoretical Chemistry*, Vol. 2, eds. R. N. Dixon and C. Thomson, p. 1, The Chemical Society, London.

G. G. Balint-Kurti (1975) in *Molecular Scattering: Physical and Chemical Applications*, ed. K. P. Lawley, p. 137, Wiley, Chichester.

G. G. Balint-Kurti and M. Karplus (1974) in *Orbital Theories of Molecules and Solids*, ed. N. H. March, p. 250, Clarendon Press, Oxford.

G. G. Balint-Kurti and R. N. Yardley (1977) *Farad. Disc. Chem. Soc.*, **62**, 77.

J. S. Binkley, J. A. Pople and W. J. Hehre (1980) *J. Amer. Chem. Soc.*, **102**, 939.

N. C. Blais and D. L. Bunker (1962) *J. Chem. Phys.*, **37**, 2713.

N. C. Blais and D. G. Truhlar (1973) *J. Chem. Phys.*, **58**, 1090.

F. W. Bobrowicz and W. A. Goddard (1977) in *Methods of Electronic Structure Theory*, ed. H. F. Schaefer, p. 79, Plenum Press, New York.

J. M. Bowman and A. Kuppermann (1975) *Chem. Phys. Lett.*, **34**, 523.

R. J. Buenker and S. D. Peyerimhoff (1974) *Theor. Chim. Acta (Berl).*, **35**, 33.

D. L. Bunker and C. A. Parr (1970) *J. Chem. Phys.*, **52**, 5700.

J. K. Cashion and D. R. Herschbach (1964) *J. Chem. Phys.*, **40**, 2358.

P. A. Christiansen, Y. S. Lee and K. S. Pitzer (1979) *J. Chem. Phys.*, **71**, 4445.

J. N. L. Connor, W. Jakubetz and J. Manz (1975) *Mol. Phys.*, **29**, 347.

C. A. Coulson (1961) *Valence*, 2nd edition, Clarendon Press, Oxford.

M. J. S. Dewar (1977) *Farad. Disc. Chem. Soc.*, **62**, 197.

R. N. Dixon and I. L. Robertson (1978) in *Theoretical Chemistry*, Vol. 3, eds. R. N. Dixon and C. Thomson, p. 100, The Chemical Society, London.

T. H. Dunning and P. J. Hay (1977) in *Methods of Electronic Structure Theory*, ed. H. F. Schaefer, p. 1, Plenum Press, New York.

F. O. Ellison (1963) *J. Amer. Chem. Soc.*, **85**, 3540.

B. C. Garrett and D. G. Truhlar (1979) *J. Amer. Chem. Soc.*, **101**, 4534.

J. Gerratt (1974) in *Theoretical Chemistry*, Vol. 1, ed. R. N. Dixon, p. 60, The Chemical Society, London.

S. K. Gray and J. S. Wright (1978) *J. Chem. Phys.*, **68**, 2002.

F. Grimaldi, A. Lecourt and C. Moser (1967) *Int. J. Quant. Chem.*, **1S**, 153.

P. C. Hariharan and J. A. Pople (1973) *Theor. Chim Acta (Berl.)*, **28**, 213.

W. J. Hehre, R. F. Stewart and J. A. Pople (1969) *J. Chem. Phys.*, **51**, 2657.

D. M. Hirst (1982) in *Dynamics of the Excited State*, ed. K. P. Lawley, p. 517, Wiley, Chichester.

H. S. Johnston (1966) *Gas Phase Reaction Rate Theory*, Ronald, New York.

L. Kahn, P. Baybutt and D. G. Truhlar (1976) *J. Chem. Phys.*, **65**, 3826.

L. R. Kahn, P. J. Hay and R. D. Cowan (1978) *J. Chem. Phys.*, **68**, 2386.

P. J. Kuntz (1979) in *Atom–Molecule Collision Theory—A Guide for the Experimentalist,* ed. R. B. Bernstein, p. 79, Plenum Press, New York.

P. J. Kuntz, E. M. Nemeth, J. C. Polanyi, S. D. Rosner and C. E. Young (1966) *J. Chem. Phys.*, **44**, 1168.

Y. S. Lee, W. C. Ermler and K. S. Pitzer (1977) *J. Chem. Phys.*, **67**, 5861.

S. M. Lin, J. G. Wharton and R. Grice (1973) *Mol. Phys.*, **26**, 317.

B. Liu (1973) *J. Chem. Phys.*, **58**, 1925.

D. R. McLaughlin and D. L. Thompson (1973) *J. Chem. Phys.*, **59**, 4393.

J. T. Muckerman (1981) in *Theoretical Chemistry,* Vol. 6A, ed. D. Henderson, p. 1, Academic Press, New York.

R. S. Mulliken and W. C. Ermler (1977) *Diatomic Molecules: Results of ab initio Calculations,* Academic Press, New York.

R. S. Mulliken and W. C. Ermler (1981) *Polyatomic Molecules: Results of ab initio Calculations,* Academic Press, New York.

J. N. Murrell, S. Carter, P. Huxley, S. C. Farantos and A. J. C. Varandas (1984) *Molecular Potential Energy Functions,* Wiley, Chichester.

K. Ohno and K. Morokuma (1982) *Quantum Chemistry Literature Data Base: Bibliography of ab initio Calculations for 1978–1980,* Elsevier, Amsterdam.

J. Olsen, D. L. Yeager and P. Jørgensen (1983) *Adv. Chem. Phys.*, **54**, 1.

R. G. Parr (1963) *The Quantum Theory of Molecular Electronic Structure,* Benjamin, New York.

H. Partridge and S. R. Langhoff (1981) *J. Chem. Phys.*, **74**, 2361.

L. Pauling and E. B. Wilson (1935) *Introduction to Quantum Mechanics,* McGraw-Hill, New York.

L. Pedersen and R. N. Porter (1967) *J. Chem. Phys.*, **47**, 4751.

M. Perić, R. J. Buenker and S. D. Peyerimhoff (1979) *Can. J. Chem.*, **57**, 2491.

R. N. Porter and M. Karplus (1964) *J. Chem. Phys.*, **40**, 1105.

W. G. Richards, T. E. H. Walker and R. K. Hinkley (1971) *A Bibliography of ab initio Molecular Wavefunctions,* Clarendon Press, Oxford.

W. G. Richards, T. E. H. Walker, L. Farnell and P. R. Scott (1974) *Bibliography of ab initio Molecular Wavefunctions. Supplement for 1970–1973,* Clarendon Press, Oxford.

W. G. Richards, P. R. Scott, E. A. Colbourn and A. F. Marchington (1978) *Bibliography of ab initio Molecular Wavefunctions. Supplement for 1974–77.* Clarendon Press, Oxford.

W. G. Richards, P. R. Scott, V. Sackwild and S. A. Robins (1981) *Bibliography of ab initio Molecular Wavefunctions. Supplement for 1978–80,* Clarendon Press, Oxford.

E. S. Rittner (1951) *J. Chem. Phys.*, **19**, 1030.

B. O. Roos and P. E. M. Siegbahn (1977) in *Methods of Electronic Structure Theory,* ed. H. F. Schaefer, p. 277, Plenum Press, New York.

B. O. Roos and P. E. M. Siegbahn (1980) *Int. J. Quant. Chem.*, **17**, 485.

C. C. J. Roothaan (1951) *Rev. Mod. Phys.*, **23**, 69.

C. C. J. Roothaan (1960) *Rev. Mod. Phys.*, **32**, 179.

S. Sakai, S. Kato, K. Morokuma and I. Kusonoki (1981) *J. Chem. Phys.*, **75**, 5398.

N. Sathyamurthy, J. W. Duff, C. Stroud and L. M. Raff (1977) *J. Chem. Phys.*, **67**, 3563.

N. Sathyamurthy and L. M. Raff (1975) *J. Chem. Phys.*, **63**, 464.

P. Saxe, H. F. Schaefer and N. C. Handy (1981) *Chem. Phys. Lett.*, **79**, 202.

H. F. Schaefer (1972) *The Electronic Structure of Atoms and Molecules: A Survey of Rigorous Quantum Mechanical Results*, Addison-Wesley, Reading, MA.

G. C. Schatz, A. F. Wagner, S. P. Walch and J. M. Bowman (1981) *J. Chem. Phys.*, **74**, 4984.

I. Shavitt (1977) in *Methods of Electronic Structure Theory*, ed. H. F. Schaefer, p. 189, Plenum Press, New York.

P. Siegbahn and B. Liu (1978) *J. Chem. Phys.*, **68**, 2457.

J. C. Slater (1963) *Quantum Theory of Molecules and Solids*, Vol. 1, *Electronic Structure of Molecules*, McGraw-Hill, New York.

R. F. Stewart (1970) *J. Chem. Phys.*, **52**, 431.

D. G. Truhlar and C. J. Horowitz (1978) *J. Chem. Phys.*, **68**, 2466.

D. G. Truhlar and R. E. Wyatt (1976) *Ann. Rev. Phys. Chem.*, **27**, 1.

D. G. Truhlar and R. E. Wyatt (1977) *Adv. Chem. Phys.*, **36**, 141.

J. C. Tully (1977) in *Semi-Empirical Methods of Electronic Structure Calculations*, part A, ed. G. A. Segal, p. 173, Plenum Press, New York.

J. C. Tully (1980) in *Potential Energy Surfaces*, ed. K. P. Lawley, p. 63, Wiley, Chichester.

A. F. Wagner, G. C. Schatz and J. M. Bowman (1981) *J. Chem. Phys.*, **74**, 4960.

A. C. Wahl and G. Das (1970) *Adv. Quant. Chem.*, **5**, 261.

A. C. Wahl and G. Das (1977) in *Methods of Electronic Structure Theory*, ed. H. F. Schaefer, p. 51, Plenum Press, New York.

F. T. Wall and R. N. Porter (1962) *J. Chem. Phys.*, **36**, 3256.

Y. Zeiri and M. Shapiro (1978) *Chem. Phys.*, **31**, 217.

CHAPTER 6

theoretical studies of chemical dynamics

6.1. Detailed Rate Constants and Reaction Cross-Sections

In this chapter we shall be concerned with the calculation of the dynamics of chemical reactions using theoretical potential energy surfaces. The kinetics and dynamics of a reaction are determined by the motion of the atoms on the potential surface and in subsequent sections we shall discuss methods for dynamical calculations, but first we must define detailed rate constants and reaction cross-sections and discuss the relationship between these observables and the motion of the atoms on a potential energy surface.

In chemical kinetics one usually measures the rate constant $k(T)$ which is temperature-dependent. For a bimolecular reaction

$$A + B \rightarrow C + D \tag{6.1}$$

the rate constant $k(T)$ relates the change in concentration of one of the reactants or products to the concentrations of the reactants, e.g.

$$-\frac{dn_A}{dt} = k(T)n_A n_B, \tag{6.2}$$

where n_A and n_B are the concentrations of reactants A and B. The reaction in equation 6.1 is really a superposition of a large number of state-to-state reactions in which species A and B in well defined initial quantum states i and j react to form products C and D which are also in well defined quantum states l and m respectively. Each of these individual state-to-state reactions will have a state-to-state or detailed rate constant $k_{ij,lm}$,

$$A(i) + B(j) \xrightarrow{\ k_{ij,lm}\ } C(l) + D(m). \tag{6.3}$$

The total rate of reaction for $A(i)$ and $B(j)$ can be written

$$-\frac{dn_{A(i)}}{dt} = -\frac{dn_{B(j)}}{dt} = \sum_l \sum_m k_{ij,lm} n_{A(i)} n_{B(j)} = k_{ij} n_{A(i)} n_{B(j)}, \tag{6.4}$$

where k_{ij} is also a detailed rate constant for which only the initial quantum states are defined. It is now possible in some cases to make kinetic studies in which some of the quantum states of the reactants are selected and to make some deductions about the quantum states of the products thus giving detailed rate constants. In order to calculate the total rate constant $k(T)$, defined in terms of concentrations of A and B, we have to relate the concentrations $n_{A(i)}$ and $n_{B(j)}$ of A and B in the quantum states i and j to the total concentrations n_A and n_B by the use of distribution functions $f_{A(i)}$ and $f_{B(j)}$ giving the fractions of A and B in states i and j. The functions $f_{A(i)}$ and $f_{B(j)}$ are usually given by the Boltzmann distribution functions. The total rate of reaction can then be written

$$-\frac{dn_A}{dt} = -\frac{dn_B}{dt} = \sum_i \sum_j k_{ij} f_{A(i)} n_A f_{B(j)} n_B \tag{6.5}$$

and thus

$$k(T) = \sum_i \sum_j f_{A(i)} f_{B(j)} k_{ij}. \tag{6.6}$$

Data obtained in molecular beam studies of chemical reactions are usually discussed in terms of reaction cross-sections. We will define total and differential cross-sections and relate these to the rate constants defined above. We start by considering a non-reactive system in which a beam of species A undergoes attenuation on passing through a collision chamber containing species B. Let $I(x)$ be the flux of species A in the x-direction. Flux is defined as the number of particles crossing a unit area perpendicular to the direction of the beam in unit time and is equal to $v n_A(x)$, where v is the velocity of the beam and $n_A(x)$ is the number density at x. The fractional loss in flux due to molecules being deflected out of the beam while traversing a distance Δx is $\Delta x / \lambda$, where λ is the mean free path. Thus

$$\frac{I(x) - I(x+\Delta x)}{I(x)} = -\frac{\Delta I(x)}{I(x)} \approx \frac{\Delta x}{\lambda} \tag{6.7}$$

and

$$-\frac{dI}{I dx} = -\frac{d(\ln I)}{dx} = \frac{1}{\lambda} \tag{6.8}$$

which can be integrated to give

$$I(x) = I(0)\exp(-x/\lambda). \tag{6.9}$$

The probability of a collision occurring within the interval Δx is proportional to $n_B(x)\Delta x$ where $n_B(x)$ is the number density of B molecules. Thus

$$\frac{\Delta x}{\lambda} \propto n_B(x)\Delta x \tag{6.10}$$

which can be rewritten as

$$\frac{1}{\lambda} = \sigma n_B, \tag{6.11}$$

where the proportionality constant σ, which has the dimensions of area, is the collision cross-section. Substitution into equation 6.9 gives

$$I(x) = I(0)\exp(-n_B\sigma x). \tag{6.12}$$

It can be readily shown that for collisions between hard spheres, the collision cross-section σ is equal to πd^2, where d is the distance between the centres of the colliding spheres. For realistic intermolecular potentials the collision cross-section is the effective or equivalent area such that $(n\sigma)^{-1}$ is equal to the mean free path. The collision cross-section can also be related to the collision number Z. The number of collisions between species A and B per unit volume per unit time is given by Zn_An_B. If we consider the number of collisions per unit volume occurring in the interval Δx, this is given by $[I(x) - I(x + \Delta x)]/\Delta x$. Therefore

$$-\frac{dI}{dx} = Zn_An_B, \tag{6.13}$$

but from equation 6.8

$$-\frac{dI}{dx} = \frac{I}{\lambda} \tag{6.14}$$

and thus

$$Z = \frac{I}{n_An_B\lambda} = v\sigma. \tag{6.15}$$

The diminution in intensity due to the scattering by collisions is thus given by

$$-\frac{dI}{dx} = v\sigma n_An_B. \tag{6.16}$$

Let us now consider a system in which a chemical reaction is taking place as a beam of species A passes through a collision cell containing species B. The diminution in intensity $-(dI/dx)_R$ due to reaction can be written, by analogy with equation 6.16, as

$$-\left(\frac{dI}{dx}\right)_R = v\sigma_R n_An_B, \tag{6.17}$$

where σ_R is the reactive cross-section. The reactive cross-section is usually smaller than the collision cross-section and is a measure of the effective size of the reacting molecules as determined by their propensity to react. Comparison with equation 6.2 indicates that the term $v\sigma_R$ in equation 6.17 is essentially the rate constant $k(v)$ for collisions occurring with a relative velocity v. In a reactive system there is a whole range of relative velocities and the reactive cross-section σ_R is velocity dependent. The thermal rate constant $k(T)$ is obtained by multiplying $v\sigma_R(v)$ (which is equivalent to $k(v)$) by $f(v; T)$, the fraction of collision partners having a relative velocity v at temperature T, and integrating with respect to v

$$k(T) = \int v\sigma_R(v)f(v;T)\,dv. \tag{6.18}$$

If the velocity distribution is Maxwell–Boltzmann, $f(v;T)$ will be the Maxwell–Boltzmann distribution function

$$f(v;T) = 4\pi\left(\frac{\mu}{2\pi k_B T}\right)^{3/2} v^2 \exp\left(-\frac{\mu v^2}{2k_B T}\right), \tag{6.19}$$

where μ is the reduced mass and k_B is the Boltzmann constant.

We can also define state-to-state reactive cross-sections $S_R(ij, lm)$ for the reaction of equation 6.3 in a manner analogous to that for state-to-state rate constants. The total reactive cross-section for a particular relative velocity $\sigma_R(v)$ is related to the state-to-state cross-sections by the equation

$$\sigma_R(v) = \sum_i \sum_j \sum_l \sum_m f_i f_j S_R(i, j, l, m). \tag{6.20}$$

For a molecule in a vibration–rotation state defined by the quantum numbers v_i and J_i, the fraction of molecules in that state is given by the product of the vibrational and rotational Boltzmann distribution functions. For a diatomic molecule this will have the form

$$f_i = \frac{\exp[-hc\omega(v_i+\tfrac{1}{2})/kT]}{Q_{vib}} \frac{g_J \exp[-hcBJ_i(J_i+1)/kT]}{Q_{rot}}, \tag{6.21}$$

where ω is the vibration wavenumber, Q_{vib} is the vibrational partition function, g_J is the degeneracy $2J+1$ of level J, B is the rotational constant and Q_{rot} is the rotational partition function.

In molecular beam studies, beams of species A and B intersect at right angles. A detector, which rotates in the plane defined by the beams, measures the intensity of a given product deflected through a particular angle Θ in the laboratory space (see figure 6.1). It is also usual to measure the velocity (or energy) distribution of the product molecules. Molecular beam data are interpreted in terms of a centre-of-mass coordinate system and all laboratory velocities and angles are transformed into velocities and angles relative to the centre of mass. The collision can then be viewed as in figure 6.2, in which particle A has a velocity equal to the relative velocity, particle B is regarded as being stationary and the perpendicular distance between the relative velocity v and particle B is known as the impact parameter b. The transformed intensity data give a quantity proportional to $dN(\theta, \phi, v)$, the number of molecules scattered per unit time into an element of solid angle defined by θ to $\theta+d\theta$ and ϕ to $\phi+d\phi$. The total number of A molecules scattered into all angles, per unit time, is given by $N(v)$, where

$$N(v) = Zn_A n_B \Delta V = n_A n_B v\sigma(v)\Delta V = IN_B\sigma(v), \tag{6.22}$$

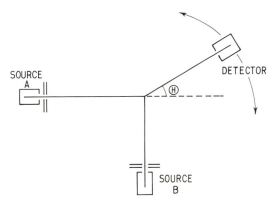

Figure 6.1. Schematic diagram of a crossed molecular beam apparatus.

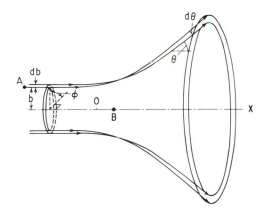

Figure 6.2. Scattering of a particle A (with effective mass μ and impact parameter b) through an angle θ by a particle B at rest.

where ΔV is the volume of the scattering region and $N_B = n_B \Delta V$ is the total number of B particles. Thus the number $dN(\theta, \phi)$ scattered into the solid angle $d\omega$, defined by the range of angles θ to $\theta + d\theta$ and ϕ to $\phi + d\phi$, must be given by an expression of the type

$$dN(\theta, \phi, v) = n_A N_B v \, d\sigma(\theta, \phi, v) = I N_B \, d\sigma(\theta, \phi, v) \qquad (6.23)$$

such that

$$N(v) = \int dN(\theta, \phi, v) \, d\omega. \qquad (6.24)$$

The quantity $d\sigma(\theta, \phi, v)$ is known as the differential cross-section and clearly satisfies the relationship

$$\sigma(v) = \int d\sigma(\theta, \phi, v) \, d\omega. \tag{6.25}$$

The differential cross-section is the number of particles scattered into the element of solid angle $d\omega$ per unit time for unit incident flux per target molecule. Differential cross-sections can be defined for both non-reactive and reactive collisions.

It can be shown that for potentials which depend only on the separation R between the two particles A and B that the scattering is cylindrically symmetrical with respect to the axis OX. All trajectories entering through an annulus of radius b and width db will be scattered into an annular cone defined by the angles θ and $d\theta$ (figure 6.2). The differential cross-section $d\sigma(\theta, \phi, v)$ is independent of ϕ. The solid angle subtended at O by the annular cone is given by $d\omega = 2\pi \sin \theta \, d\theta$ and equation 6.25 can be rewritten

$$\sigma(v) = 2\pi \int d\sigma'(\theta, v) \sin \theta \, d\theta, \tag{6.26}$$

where $d\sigma'(\theta, v)$ is now the differential cross-section for particles being scattered through an angle θ. The differential cross-section $d\sigma'(\theta, v)$ can be obtained by transforming the laboratory intensities, which are functions of angle and velocity, to the centre-of-mass coordinate system. One can, of course, also define state-to-state differential cross-sections analogous to the state-to-state total cross-sections of equation 6.20.

It was stated above that in figure 6.2 all trajectories entering through an annulus of radius b and width db will result in scattering into an annular cone defined by the angle θ. The cross-sectional area of the annulus is $2\pi b \, db$ and the collision cross-section, which is the effective area perpendicular to v such that, for a collision to take place, the trajectory must be within the area, can be written

$$\sigma(v) = \int_0^{b_{max}} 2\pi b \, db, \tag{6.27}$$

where b_{max} is the maximum value of b for which scattering (or reaction) takes place. For hard spheres, b_{max} is simply equal to d, the sum of the radii and the collision cross-section is equal to πd^2. Thus from equation 6.25 we see that

$$d\sigma(\theta, \phi, v) \, d\omega = 2\pi b \, db \tag{6.28}$$

or

$$d\sigma'(\theta, v) 2\pi \sin \theta \, d\theta = 2\pi b \, db. \tag{6.29}$$

Thus

$$d\sigma'(\theta, v) \sin \theta \, d\theta = b \, db. \tag{6.30}$$

In the case of reactive collisions, not every collision will result in reaction and the term $2\pi b \ db$ in equation 6.27 should be multiplied by a factor $P(b)$ which is the fraction of collisions with impact parameter b leading to reaction

$$\sigma_R(v) = \int_0^{b_{\text{max}}} P(b)2\pi b \ db. \tag{6.31}$$

The function $P(b)$ is known as the opacity function. Provided $P(b)$ does not fall off too rapidly with increasing b, the factor $2\pi b$ means that collisions with larger values of b make an increasing contribution to the reactive cross-section. Thus if $P(b)$ extends to fairly large values of b, the reaction will have a large cross-section.

Another useful parameter is obtained by expressing the reactive cross-section as a function of the initial relative energy rather than the velocity. This function $\sigma_R(E)$ is known as the excitation function.

6.2. Classical Trajectory Calculations

For a rigorous description of the dynamics of a collision, it is necessary to use quantum mechanics. However, atoms are much heavier than electrons and in many circumstances it is found that classical mechanics can give a perfectly satisfactory description of the motion of atoms on a potential surface. By solving classical equations of motion one can calculate the trajectories of the atoms (or molecules) for a given set of starting conditions. Since it is not possible to define experimental conditions so precisely, typically a large number of calculations are performed for different sets of initial conditions, and by suitable averaging, cross-sections and rate constants can be derived.

There are, however, several situations in which a classical treatment will be inadequate and for which a quantum mechanical or semi-classical approach should be used. Classical mechanics cannot be expected to be satisfactory when the total energy available for the reaction is comparable with or less than the energy of the potential barrier separating products from reactants. In this regime, quantum mechanical tunnelling may be significant and classical methods will underestimate the probability of reaction. Also classical methods are not suitable in situations where resonance or interference effects would be expected to be important. In many situations, however, quantum effects would be expected to be unimportant. One would expect classical methods to be more reliable as the masses of the atoms increase, because the de Broglie wavelength will be smaller. Similarly the method should be more reliable for higher energies in each degree of freedom. As indicated above, it is necessary to average over many trajectories in order to compare the results of the calculations with experiment and this averaging process also increases the accuracy. We restrict ourselves here to the adiabatic case, where the reaction can be discussed in terms of motion on a single potential energy surface. We defer discussion of the non-adiabatic case to section 6.4.

Quantum mechanical and semi-classical methods are beyond the scope of this book and for discussion of these methods the reader is referred to Child (1974, 1976), Miller (1976), Connor (1979), Bernstein (1979) and Kuppermann (1981).

6.2.1. *Hamilton's Equations of Motion for Trajectory Calculations*

The dynamical problem is usually formulated in terms of Hamilton's equations of motion (for a full discussion see Goldstein 1950), in which derivatives, with respect to time, of the coordinates and components of the linear momentum of each particle are expressed as partial derivatives of the Hamiltonian function H. The Hamiltonian function, in the absence of magnetic fields, is simply the classical expression for the total energy written as the sum of the kinetic energy T, in terms of momenta, and the potential energy V.

$$H = T + V. \tag{6.32}$$

We begin with a discussion in terms of space fixed axes x, y, z. For three atoms A, B, C with masses m_A, m_B, m_C, the kinetic energy is given in this coordinate system by

$$T = \sum_{i=x,y,z} \left(\frac{1}{2m_A} p_{A_i}^2 + \frac{1}{2m_B} p_{B_i}^2 + \frac{1}{2m_C} p_{C_i}^2 \right), \tag{6.33}$$

where p_{A_i} etc. are the components of the momentum. The potential energy V will be a function of the positions of the particles

$$V = V(x_A, y_A, z_A, x_B, \dots). \tag{6.34}$$

With this definition of the Hamiltonian function, Hamilton's equations are of the form

$$\dot{x}_A = \frac{\partial H}{\partial p_{A_x}}, \tag{6.35}$$

$$\dot{p}_{A_x} = -\frac{\partial H}{\partial x_A}, \tag{6.36}$$

where $\dot{x}_A$ is the derivative dx_A/dt. Thus there will be a total of 18 equations of which 9 relate to derivatives of the position coordinates and 9 to derivatives of the momentum coordinates.

The advantage of the Hamiltonian formulation is that the form of equations 6.35 and 6.36 remains unchanged when the coordinate system is transformed from the space fixed system (x_A, y_A, z_A, etc.) to a set of generalized coordinates ($q_1, q_2, \ldots, q_9$) which may be more suitable for the problem in hand. The corresponding generalized momentum p_i is defined in terms of q_i and the Lagrangian function L

$$L = T - V \tag{6.37}$$

by

$$p_i = \frac{\partial L}{\partial \dot{q}_i}. \tag{6.38}$$

With this definition of the generalized momentum, Hamilton's equations become

$$\dot{q}_i = \frac{\partial H}{\partial p_i}, \tag{6.39}$$

$$\dot{p}_i = -\frac{\partial H}{\partial q_i}. \tag{6.40}$$

In the problem of the dynamics of the three particles A, B and C we do not need to consider the motion of the centre of mass of the system provided that there are no external forces. Thus we can choose a set of nine generalized coordinates such that three of them represent the position of the centre of mass. This will enable us to reduce the number of equations to be solved from 18 to 12. For the collision of A with BC this can be achieved by the following transformations

$$Q_1 = x_C - x_B, \tag{6.41}$$

$$Q_4 = x_A - \frac{m_B x_B + m_C x_C}{m_B + m_C} \tag{6.42}$$

and

$$Q_7 = \frac{m_A x_A + m_B x_B + m_C x_C}{M}, \tag{6.43}$$

with corresponding expressions Q_2, Q_5 and Q_8 for the y-axis and Q_3, Q_6 and Q_9 for the z-axis. M is the total mass $m_A + m_B + m_C$. Q_7, Q_8, Q_9 are the coordinates of the centre of mass of the system. Q_1, Q_2, Q_3 give the coordinates of C relative to B and Q_4, Q_5, Q_6 are the coordinates of A relative to the centre of mass of BC. The reverse transformation is given by the equations

$$x_A = \frac{m_B + m_C}{M} Q_4 + Q_7, \tag{6.44}$$

$$x_B = -\frac{m_C}{m_B + m_C} Q_1 - \frac{m_A}{M} Q_4 + Q_7, \tag{6.45}$$

$$x_C = \frac{m_B}{m_B + m_C} Q_1 - \frac{m_A}{M} Q_4 + Q_7, \tag{6.46}$$

etc. The momenta p_i in the space fixed axis coordinate system are related to the generalized momenta P_i by the relationship

$$p_i = \sum_j P_j \frac{\partial Q_j}{\partial q_i}, \tag{6.47}$$

giving the expressions

$$p_{x_A} = P_4 + \frac{m_A}{M} P_7, \tag{6.48}$$

$$p_{x_B} = -P_1 - \frac{m_B}{m_B + m_C} P_4 + \frac{m_B}{M} P_7, \tag{6.49}$$

$$p_{x_C} = P_1 - \frac{m_C}{m_B + m_C} P_4 + \frac{m_C}{M} P_7, \tag{6.50}$$

etc. The Hamiltonian function is obtained by substituting the expressions of equations 6.48 to 6.50 into equation 6.33 for the kinetic energy and expressing the potential energy V in terms of the generalized coordinates Q_i to give

$$H = \frac{1}{2\mu_{BC}} \sum_{i=1}^{3} P_i^2 + \frac{1}{2\mu_{A,BC}} \sum_{i=4}^{6} P_i^2$$
$$+ \frac{1}{2M} \sum_{i=7}^{9} P_i^2 + V(Q_1, \dots, Q_6), \tag{6.51}$$

where the reduced masses μ_{BC} and $\mu_{A,BC}$ are given by

$$\frac{1}{\mu_{BC}} = \frac{1}{m_B} + \frac{1}{m_C}; \qquad \frac{1}{\mu_{A,BC}} = \frac{1}{m_A} + \frac{1}{m_B + m_C}. \tag{6.52}$$

The potential energy is independent of the position of the centre of mass and thus involves only the generalized coordinates Q_1 to Q_6. The potential energy function is usually expressed in terms of the interatomic distances R_{AB}, R_{BC} and R_{AC} and it is therefore necessary to write these in terms of the x, y, z coordinates of each atom, e.g.

$$R_{AB} = [(x_B - x_A)^2 + (y_B - y_A)^2 + (z_B - z_A)^2]^{\frac{1}{2}}$$

$$= \left[\left(\frac{m_C}{m_B + m_C} Q_1 + Q_4 \right)^2 + \left(\frac{m_C}{m_B + m_C} Q_2 + Q_5 \right)^2 \right.$$

$$\left. + \left(\frac{m_C}{m_B + m_C} Q_3 + Q_6 \right)^2 \right]^{\frac{1}{2}}. \tag{6.53}$$

The potential energy function is transformed from a function in terms of R_{AB}, R_{BC} and R_{AC} to $V(Q_1, \ldots, Q_6)$ by use of the chain rule

$$\frac{\partial V}{\partial Q_i} = \sum_j \frac{\partial V}{\partial R_j} \frac{\partial R_j}{\partial Q_i}, \tag{6.54}$$

where $R_j = R_{AB}, R_{BC}, R_{AC}$.

The Hamilton equations for $\dot{Q}_i$ assume very simple forms, namely

$$\dot{Q}_i = \frac{1}{\mu_{BC}} P_i \quad (i = 1, 2, 3), \tag{6.55}$$

$$\dot{Q}_i = \frac{1}{\mu_{A,BC}} P_i \quad (i = 4, 5, 6), \tag{6.56}$$

$$\dot{Q}_i = \frac{1}{M} P_i \quad (i = 7, 8, 9). \tag{6.57}$$

The equations for $\dot{P}_i$ are as follows

$$-\dot{P}_i = \frac{1}{R_{AB}} \frac{m_C}{m_B + m_C} \left(\frac{m_C}{m_B + m_C} Q_i + Q_{i+3} \right) \frac{\partial V}{\partial R_{AB}} + \frac{Q_i}{R_{BC}} \frac{\partial V}{\partial R_{BC}}$$

$$+ \frac{1}{R_{AC}} \frac{m_B}{m_B + m_C} \left(\frac{m_B}{m_B + m_C} Q_i - Q_{i+3} \right) \frac{\partial V}{\partial R_{AC}} \quad (i = 1, 2, 3) \tag{6.58}$$

$$-\dot{P}_i = \frac{1}{R_{AB}} \left(\frac{m_C}{m_B + m_C} Q_i + Q_{i+3} \right) \frac{\partial V}{\partial R_{AB}}$$

$$- \frac{1}{R_{AC}} \left(\frac{m_B}{m_B + m_C} Q_i - Q_{i+3} \right) \frac{\partial V}{\partial R_{AC}} \quad (i = 4, 5, 6) \tag{6.59}$$

$$\dot{P}_i = 0 \quad (i = 7, 8, 9). \tag{6.60}$$

The momenta P_7, P_8 and P_9 are constants of the motion and we only need to consider the equations for $i = 1$ to 6.

Trajectories (i.e. the values of the coordinates Q_i as functions of time) are calculated by numerical integration of Hamilton's equations on a computer. The integration is done in a stepwise manner in which the right-hand sides of equations 6.55, 6.56, 6.58 and 6.59 are evaluated for some value of the time t. From the values of the derivatives $\dot{Q}_i$ and $\dot{P}_i$ the values of Q_i and

P_i are calculated for a later time $t + \Delta t$ and repetition of this process yields Q_i and P_i as functions of time. We will not be concerned here with details of the integration process, for which sophisticated computer routines are available which enable the trajectories to be calculated very rapidly. For details the reader is referred to reviews by Bunker (1971), Porter and Raff (1976) and Truhlar and Muckerman (1979). A set of initial conditions is specified and usually integration is continued until the separation between two of the atoms is so large that there is no longer any interaction between them. The accuracy of the integration should be monitored by checking that the total energy or total angular momentum is conserved. A more rigorous test is 'back' integration from the final state to the initial state. The selection of initial conditions and of the final state of the system will be discussed below.

6.2.2. Selection of Initial Conditions and the Monte Carlo Method

Before starting the numerical integration it is necessary to specify initial values for the 12 coordinates Q_i^0 and P_i^0 ($i = 1$ to 6). We can, without loss of generality, choose a coordinate system in which the initial relative velocity vector V_R lies along the z-axis and the centre of mass of BC lies in the yz plane (see figure 6.3).

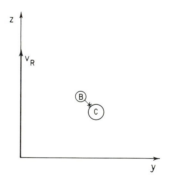

Figure 6.3. Coordinate system for initial conditions in a quasi-classical trajectory calculation.

The initial conditions clearly include the initial relative velocity V_R and the distance ρ between atom A and the centre of mass of BC. In addition, it is necessary to specify the impact parameter b which is the perpendicular distance between the centre of mass of BC and the initial relative velocity vector V_R. With the coordinate system defined above we obtain the following values for coordinates 4 to 6 which are concerned with the motion of A relative to BC.

$$Q_4^0 = 0; \qquad Q_5^0 = b; \qquad Q_6^0 = -(\rho^2 - b^2)^{1/2}; \tag{6.61}$$

$$P_4^0 = P_5^0 = 0; \qquad P_6^0 = \mu_{A,BC} V_R. \tag{6.62}$$

We now have to consider the internal motion of the molecule BC. It is usual to choose internal conditions such that BC has quantized rotational and vibrational energies given by quantum numbers J and v. When this is done the calculations are referred to as quasi-classical trajectory calculations. The specification of the initial conditions for BC requires two angles θ and ϕ. The angle θ is the angle between the direction of the BC bond and the z-axis and ϕ is the angle between the projection of the BC bond on to the xy plane and the x-axis. Thus

$$Q_1^0 = r^0 \sin\theta \cos\phi; \qquad Q_2^0 = r^0 \sin\theta \sin\phi; \qquad Q_3^0 = r^0 \cos\theta, \tag{6.63}$$

where r^0 is the initial internuclear separation. We also have to specify the initial orientation η of the angular momentum vector associated with the rotational motion of BC and the initial phase ξ of the vibrational motion. One way of incorporating the initial phase ξ of the vibrational motion is to integrate just the equations describing the motion of the diatomic BC for a time $\xi\tau/2\pi$ (where τ is the vibration period) from $r = r_-$, the minimum value of the BC bond length. When this time has elapsed, the integration of the equations describing the three-body motion is started. If $\xi \geqslant \pi$, then it is more efficient to integrate the diatomic equations for a time $(\xi - \pi)\tau/2\pi$ from r_+, the maximum value of the BC bond length. The initial values of P_1^0, P_2^0, P_3^0 are chosen as follows

$$P_1^0 = J_r(\sin\theta \cos\eta - \cos\theta \cos\phi \sin\eta)/r_\pm,$$

$$P_2^0 = -J_r(\cos\phi \cos\eta + \cos\theta \sin\phi \sin\eta)/r_\pm, \tag{6.64}$$

$$P_3^0 = J_r(\sin\theta \sin\eta)/r_\pm,$$

where J_r is the magnitude of the total internal angular momentum and is given by $(h/2\pi)[j(j+1)]^{1/2}$, η is the angle between the BC rotational angular momentum vector and a reference vector normal to the BC internuclear axis defined by $r \times \hat{e}_z$, where r is a vector along the internuclear axis and $\hat{e}_z$ is a unit vector along the space fixed z-axis.

Thus for a given trajectory calculation in which BC is initially in the vibration–rotation state defined by quantum numbers v,J, one has to choose values for V_R, b, ρ, θ, ϕ, η and ξ. This is done in such a way that averaging over many trajectories can be carried out to yield cross-sections and rate constants. The value of the initial separation between A and the centre of mass of BC, ρ, is not usually varied but an initial value is chosen such that there is no interaction between A and BC. For the purposes of comparison with experiment and for the calculation of reactive cross-sections one is concerned with the probability of a reaction occuring for a given value of the initial relative velocity V_R, with BC in the vibration–rotation state v, J and a given impact parameter b. This can be obtained from the trajectory calculations by the relation

$$P_R(V_R, v, J, b) = \lim_{N \to \infty} \frac{N_R(V_R, v, J, b)}{N(V_R, v, J, b)}, \tag{6.65}$$

where N_R is the number of trajectories resulting in reaction and N is the total number of trajectories, provided that values of the variables θ, ϕ, η and ξ have been properly chosen. The function $P_R(V_R, v, J, b)$ is, of course, the opacity function defined in equation 6.31. The reaction cross-section, for initial state v, J and relative velocity V_R, is obtained from P_R by integration over the impact parameter

$$S_R(V_R, v, J) = 2\pi \int_0^{b_{max}} P_R(V_R, v, J, b) b \ db, \tag{6.66}$$

where b_{max} is the maximum value of b for which reaction takes place. A value for b_{max} is chosen on the basis of a set of preliminary calculations. Calculation of the total cross-section requires further averaging over V_R and the vibration–rotation states.

In the computation of P_R in equation 6.65, averaging has been carried out with respect to the variables θ, ϕ, η and ξ. This is done by the Monte Carlo procedure for the evaluation of a multidimensional integral. The integral

$$I = \int_0^1 \int_0^1 \cdots \int_0^1 f(x_1, x_2, \dots, x_n) \ dx_1 \ dx_2 \dots dx_n \tag{6.67}$$

is approximated by taking the average value of the function $f(x_1, x_2, \dots, x_n)$ for N randomly selected sets of variables $\{x_1^{(i)}, x_2^{(i)}, \dots, x_n^{(i)}\}$, where i runs from 1 to N:

$$I \approx \frac{1}{N} \sum_{i=1}^{N} f(x_1^{(i)}, x_2^{(i)}, \dots, x_n^{(i)}). \tag{6.68}$$

It can be shown that the error is proportional to $N^{-1/2}$ so initial convergence is quite rapid and a reasonably accurate estimate can be obtained with relatively few points. The range of integration in equation 6.67 is from 0 to 1 and the variables $\{x_j^{(i)}\}$ are chosen as sets of random numbers in the range 0 to 1. One has to consider carefully the relationship between the variables θ, ϕ, η, ξ and b with respect to which integration is being carried out in equation 6.66 and the differential element dx_j in equation 6.67. If we consider the orientation of the molecule BC, there is equal probability of it having an orientation with each differential element of solid angle $d\omega = \sin\theta \ d\theta \ d\phi$. Thus for the variation of ϕ, $dx_j = c_j \ d\phi$ but for θ, $dx_k = c_k \sin\theta \ d\theta$, where c_j and c_k are proportionality constants such that x_j is unity when the variable to which it is related is at the upper end of its range. Integration of dx_k yields

$$x_k = \int_0^{x_k} dx_k = c_k \int_0^{\theta} \sin\theta \, d\theta = c_k(1-\cos\theta). \tag{6.69}$$

Since θ varies from 0 to π, c_k has the value 0.5 and

$$x_k = \tfrac{1}{2}(1-\cos\theta). \tag{6.70}$$

The variables η and ξ vary uniformly from 0 to 2π but the differential element for the impact parameter is $b \, db$. For a given trajectory calculation the initial values of θ, ϕ, η, ξ and b are obtained by taking a set of random numbers x_i, $i = 1$ to 5, in the range 0 to 1 and using the relationships

$$x_1 = \tfrac{1}{2}(1-\cos\theta); \quad x_2 = \frac{1}{2\pi}\phi; \quad x_3 = \frac{1}{2\pi}\eta;$$

$$x_4 = \frac{1}{2\pi}\xi; \quad x_5 = \left(\frac{b}{b_{\max}}\right)^2. \tag{6.71}$$

Subsequent trajectories use initial values calculated from new sets of random numbers x_i and the appropriate integration is performed using equation 6.68.

In order to calculate rate constants for state-selected reactants it is also necessary to average over the initial relative velocity V_R and Monte Carlo methods are available for this. Finally, to obtain thermal rate constants it is necessary to run batches of trajectories for various vibration–rotation states v, J of BC and include the appropriate Boltzmann weighting factor.

6.2.3. *Analysis of the Results of Trajectory Calculations*

As mentioned above, integration of Hamilton's equations is continued until the separation between two atoms is so large that there is no interaction between them. However, the trajectory should not be stopped if the two atoms in question are moving towards each other. At the end of the calculation it is necessary to analyse the trajectory to determine what the product molecule is, whether it is bound, quasi-bound or in a dissociative state, to determine the final relative kinetic energy, to partition the internal energy into vibrational and rotational energy and to obtain the scattering angle. Because the trajectories are calculated by classical mechanics, the internal energy of the product does not correspond to a particular vibration–rotation energy level. There are various methods for assigning rotational and vibrational quantum numbers for the product molecule and we refer the reader to more comprehensive discussions for details (Karplus, Porter and Sharma 1965, Polanyi and Schreiber 1974, Porter and Raff 1976, Truhlar and Muckerman 1979). The trajectories are then classified by counting the number of trajectories for which the final states outlined above fall within certain ranges to yield probability functions analogous to the function $P_R(V_R, v, J, b)$ defined in equation 6.65 for a given set of initial conditions. For example, if we are interested in the angular distribution of the product molecule in the final state v', J' resulting from the initial condition v, J, b, we will calculate the probability function

$$P_R(V, v, J, b; \Theta, v', J') = \lim_{N \to \infty} \frac{N_R(V, v, J, b; \Theta, v', J')}{N(V, v, J, b)}, \qquad (6.72)$$

where Θ is the scattering angle. These probabilities can then be used in the calculation of differential and total cross-sections, rate constants, the opacity function or the excitation function.

6.3. The Relationship Between Reaction Dynamics and the Potential Surface

There are two approaches to the application of the quasi-classical trajectory calculations outlined in the previous section. One approach seeks to reproduce as closely as possible all the known experimental data for a given reactive system. In this approach much of the effort goes into finding an appropriate potential surface and adjusting it in such a way as to bring the calculated dynamics into agreement with experiment. The other approach regards the quasi-classical trajectory method as an 'experimental' technique in which one can use several model potential surfaces and vary the initial conditions to investigate the correlation between dynamical behaviour and the properties of the potential surfaces. Although many of the calculations done with this philosophy do not directly correspond to particular reactive systems, a number of generalizations can be made which enable one to interpret particular aspects of the observed dynamics of a chemical reaction in terms of properties of the potential surface. We start our discussion with this latter approach and then in section 6.3.5 we consider in some detail the results of trajectory calculations for the system

$$F + H_2 \to FH + H \qquad (6.73)$$

which has been investigated by many workers.

The methods discussed in section 6.2 are applicable to the complete three-dimensional dynamics of a reaction of the type

$$A + BC \to AB + C. \qquad (6.74)$$

However, useful generalizations can often be derived from two-dimensional collinear calculations. In these calculations the potential surface can be represented by a single contour diagram and it is only necessary to consider translational and vibrational motion. Such calculations do not include rotational motion and cannot give any indication of the extent to which rotational excitation occurs. In many cases the collinear calculations provide a useful guide to the full three-dimensional calculation. We outline the collinear approach in the next section and in subsequent sections discuss some general conclusions with regard to the dynamics of reaction 6.74, assuming that the reaction is direct. By direct we mean that once the products start to separate, they continue to do so and that the separation between them does not subsequently start to decrease. Thus we will be concerned with situations in which there are no secondary encounters.

6.3.1. Collinear Trajectory Calculations; Skewed Coordinate Systems

If we restrict ourselves to collinear geometries for reaction 6.74, it is necessary to specify only two geometrical parameters, namely r_1, the distance between A and B, and r_2, the separation between B and C. The distance between A and C is clearly equal to $r_1 + r_2$. With this constraint the potential function depends only on r_1 and r_2 and its functional form is obtained by substituting $r_1 + r_2$ for r_{AC}. The equations of motion can be expressed simply in terms of r_1 and r_2 as

$$\frac{d^2 r_1}{dt^2} = -\frac{m_A + m_B}{m_A m_B} \left(\frac{\partial U}{\partial r_1}\right) + \frac{1}{m_B}\left(\frac{\partial U}{\partial r_2}\right), \tag{6.75}$$

$$\frac{d^2 r_2}{dt^2} = \frac{1}{m_B}\left(\frac{\partial U}{\partial r_1}\right) - \frac{m_B + m_C}{m_B m_C}\left(\frac{\partial U}{\partial r_2}\right) \tag{6.76}$$

and the kinetic energy is given by

$$T = \frac{m_A(m_B + m_C)}{2M}\dot{r}_1^2 + \frac{m_A m_C}{M}\dot{r}_1\dot{r}_2 + \frac{m_C(m_A + m_B)}{2M}\dot{r}_2^2, \tag{6.77}$$

where M is the total mass. In the kinetic energy expression there is a term in $\dot{r}_1\dot{r}_2$ which indicates coupling between motion with respect to r_1 and r_2. Similarly in the equations of motion the terms $(\partial U/\partial r_2)/m_B$ and $(\partial U/\partial r_1)/m_B$ correspond to coupling between the two coordinates. This inertial coupling will occur irrespective of the presence of $r_1 r_2$ terms in the potential energy function. Equation 6.77 is analogous to equation 4.18 encountered in the theory of molecular vibrations and we therefore expect to be able to find a new set of coordinates Q_i such that the kinetic energy can be written

$$T = \frac{1}{2M}(\dot{Q}_1^2 + \dot{Q}_2^2). \tag{6.78}$$

The advantage of this transformation is that we can now visualize the dynamics in terms of a single hypothetical particle of mass M moving in two dimensions subject to the potential $U(Q_1, Q_2)$. The trajectory can be discussed in terms of a mass sliding over the transformed potential surface and can be represented diagrammatically on the contour diagram of the transformed potential surface.

The transformation of the kinetic energy from the representation of equation 6.77 to that of equation 6.78 is effected by skewing and scaling the axes r_1, r_2 in the two-dimensional representation of the potential surface by writing

$$Q_1 = r_1 + \beta r_2 \sin \phi, \tag{6.79}$$

$$Q_2 = \beta r_2 \cos \phi. \tag{6.80}$$

The reverse transformation is

$$r_1 = Q_1 - Q_2 \tan \phi, \tag{6.81}$$

$$r_2 = \frac{Q_2}{\beta} \sec \phi. \tag{6.82}$$

The parameters β and ϕ depend on the masses of the three atoms and are given by

$$\sin \phi = \left(\frac{m_A m_C}{(m_A + m_B)(m_B + m_C)} \right)^{1/2}, \tag{6.83}$$

$$\beta = \left(\frac{m_C(m_A + m_B)}{m_A(m_B + m_C)} \right)^{1/2}. \tag{6.84}$$

This is illustrated in figure 6.4 where it can be seen that if $\beta > 1$, the potential surface is extended along the βr_2 axis (the exit valley), whereas if $\beta < 1$ it is compressed along this axis. Kuntz (1976) gives a table of values of ϕ and β for all the possible mass combinations of light (L) and heavy (H) atoms with $m_L = 1$ u and $m_H = 80$ u. In the new coordinate system, the two axes are drawn at an angle of $90° - \phi$ with respect to each other. The potential function is plotted as a function of r_1 and r_2 but with the units on the r_2 axis being scaled by the factor β.

The transformation parameters ϕ and β can be interpreted as follows. Inertial coupling between motion in r_1 and r_2 is small if ϕ is small, which is the case when the mass of A or of C is much smaller than that of B. However if ϕ is large, there is strong coupling between the two coordinates. The relative time scales for motion with respect to r_1 and r_2 are related to the parameter β. The exit valley is extended if $\beta > 1$ and the motion in the entrance valley is faster for equal forces.

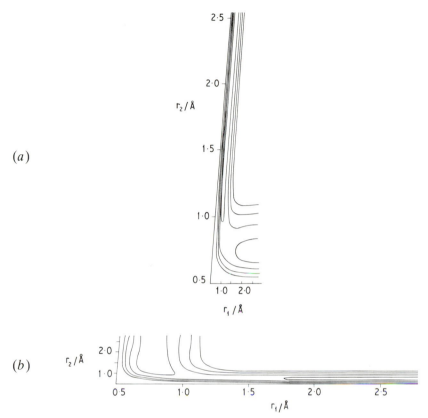

Figure 6.4. Skewed and scaled potential energy surfaces for mass combinations (*a*) L + HH; (*b*) H + HL.

6.3.2. *Attractive, Mixed and Repulsive Energy Release*

For exothermic reactions the manner in which energy is released as the system moves from the barrier on the surface down to the products has a considerable influence on the dynamics of the reaction. In the reactions of alkali metals M with halogens XY, there is an avoided crossing in the entrance channel between the covalent potential surface M + XY and the ionic surface $M^+ + XY^-$ (arising from considerations similar to those discussed in the case of NaCl in section 2.7). Once the avoided crossing has been passed, the surface is basically ionic and there is a strong attractive force between the reactants. It had been suggested on the basis of early work on these systems that the strong force of attraction should give rise to vibrational excitation of the products.

Polanyi and his co-workers (see Kuntz *et al.* 1966) have given several definitions for the attractive and repulsive character of potential surfaces. In

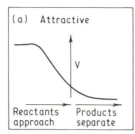 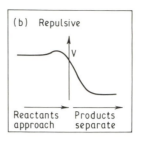

Figure 6.5. Schematic cuts through potential surfaces illustrating (*a*) attractive energy release; (*b*) repulsive energy release. The vertical line represents the geometry for which the ratio r_{AB}/r_{BC} is equal to the ratio of the equilibrium values $(r_{AB})_e/(r_{BC})_e$.

qualitative terms attractive energy release is the energy released as the reactants approach and the bond between A and B is being formed, whereas repulsive energy release is that released as the products separate and the BC bond is broken. This is illustrated in figure 6.5, where the vertical line represents the geometry for which $r_{AB}/r_{BC} = (r_{AB})_e/(r_{BC})_e$, where $(r_{AB})_e$ represents the equilibrium value of r_{AB}.

In figure 6.5 (*a*) most of the energy is released before this point and this situation corresponds to attractive energy release, whereas in figure 6.5 (*b*) most of the energy is released after this point and we have repulsive energy release. A potential surface of the type illustrated in figure 6.5 (*a*) is sometimes referred to as an early downhill surface and a repulsive surface (figure 6.5 (*b*)) as a late downhill surface.

We consider substantially exothermic reactions where ΔH is of the order of 125–200 kJ mol^{-1}, for which the activation barrier is relatively small compared with the exothermicity. Thus the activation barrier is not a substantial feature of the potential surface. The simplest classification is the rectilinear classification (Kuntz *et al.* 1966) which divides the energy release into attractive $A_\perp$ and repulsive $R_\perp$ energy release. The energy released on going from the top of the barrier to products is $E_c - \Delta H$, where E_c is the height of the barrier relative to reactants and ΔH is the exothermicity of the reaction. The attractive energy release $A_\perp$ is defined as the percentage of the energy $E_c - \Delta H$ released as A approaches BC collinearly (with BC 'clamped' at its normal equilibrium distance r_2^0) from the barrier up to the point where r_1 equals the normal AB distance r_1^0 or the point at which repulsion between A and BC starts to occur. The repulsive component is then simply $100 - A_\perp$. Three potential surfaces with values of $A_\perp$ of 4%, 47% and 72% are illustrated in figure 6.6.

Two-dimensional trajectory calculations for a light atom L reacting with a diatomic HH containing two heavy atoms have been made for a series

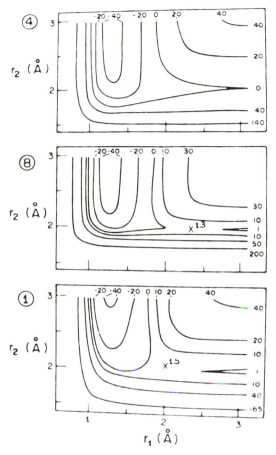

Figure 6.6. Collinear potential surfaces for which the attractive energy release $A_\perp$ is (a) 4%; (b) 47%; (c) 72%. (Reproduced from Kuntz *et al.* 1966.)

of extended LEPS surfaces (see section 5.2.1) for which $A_\perp$ varies from 4% to 72%. These calculations indicate that the mean vibrational excitation increases systematically as $A_\perp$ increases. Comparable three-dimensional studies show similar trends. The L+HH case is not representative because a light atom A can approach the heavy atom B in BC to a distance which is comparable with the normal AB bond length before separation of B and C starts to occur. Repulsion between B and C is thus not efficient in exciting the AB vibration. Such a strong correlation between $A_\perp$ and vibrational excitation is not found for other mass combinations where substantial vibrational excitation is obtained for surfaces with low values of $A_\perp$. The atypical behaviour of the L+HH system on repulsive potential surfaces is known as the 'light atom anomaly'.

The division of energy into attractive and repulsive energy release is an over-simplification and there are two other schemes for partitioning the energy release. A third category of energy release, called mixed energy release, is introduced to describe the situation while the trajectory is 'cutting the corner' on the potential energy surface. Usually atom A will be approaching B while BC repulsion energy is being released. While this is occurring r_1 is decreasing and r_2 is increasing. In terms of a sliding mass on a two-dimensional surface, this part of the trajectory brings the particle on to the side of the exit channel thus, resulting in vibration of the product.

One classification scheme (Kuntz *et al.* 1966) considers motion along the minimum energy pathway over the surface which goes asymptotically to r_2^0 (for A + BC) and to r_1^0 (for AB + C) (see figure 6.7).

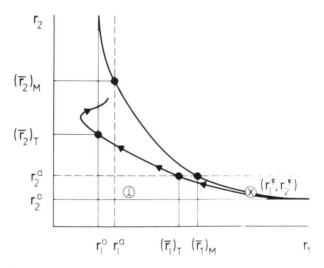

Figure 6.7. Minimum energy pathway and trajectory classification of attractive, mixed and repulsive energy release. (Reproduced from Kuntz 1976.)

The position of the barrier is given by the coordinates $(r_1^\ddagger, r_2^\ddagger)$. We require the definition of additional values of r_1 and r_2, namely r_1^a and r_2^a. These are estimates of the maximum extensions of the AB and BC bonds during vibration. Attractive energy release is regarded as occurring as the particle moves along the minimum energy pathway from the barrier at $(r_1^\ddagger, r_2^\ddagger)$ to a point where r_2 has increased to r_2^a. At this point r_1 has a value $(\bar{r}_1)_M$. Mixed energy release occurs as r_1 decreases from this value to r_1^a while r_2 increases to $(\bar{r}_2)_M$. Repulsive energy release then occurs as r_2 increases from $(\bar{r}_2)_M$ to infinity with r_1 decreasing to r_1^0. Thus we obtain the following expressions for attractive energy release A_M, mixed energy release M_M and repulsive energy release R_M:

$$A_M = \frac{100[U(r_1^{\ddagger}, r_2^{\ddagger}) - U((\bar{r}_1)_M, r_2^a)]}{E_c - \Delta H}$$

$$= 0 \quad \text{if } (\bar{r}_1)_M > r_1^{\ddagger}, \tag{6.85}$$

$$M_M = \frac{100[U((\bar{r}_1)_M, r_2^a) - U(r_1^a, (\bar{r}_2)_M)]}{E_c - \Delta H}$$

$$= 0 \quad \text{if } r_1^a > (\bar{r}_1)_M, \tag{6.86}$$

$$R_M = 100 - A_M - M_M. \tag{6.87}$$

This method is unsatisfactory when various mass combinations are considered because the characteristic path across the surface will depend on the particular mass combination. In such cases one should use a trajectory classification based on a collinear trajectory with an initial energy equal to the average thermal energy (if there is no barrier) or the threshold energy for reactions with a barrier. Attractive energy release A_T occurs along the portion of the trajectory from $(r_1^{\ddagger}, r_2^{\ddagger})$ to the point at which it cuts the line $r_2 = r_2^a$. Mixed energy release M_T takes place from this point until $r_1 = r_1^0$ and repulsive energy release R_T over the remainder of the trajectory. Parameters A_T, M_T and R_T are defined by equations analogous to equations 6.85 to 6.87. Mixed energy release is channelled into vibrational energy and the mean vibrational excitation shows a good correlation with the sum $A_T + M_T$ for all mass combinations.

Mixed energy release is not possible for the mass combination $L + HH$. This can be understood in terms of the skewed and scaled potential surface obtained by applying the transformation of equations 6.81 and 6.82 to a repulsive potential surface. The parameter β has a very large value, resulting in a skewed potential surface in which the entrance channel is very wide and flat and the exit channel is very long and narrow (see figure 6.8). Thus in order to move down the exit channel, the particle must start at its head and consequently will acquire little vibrational energy. However if we consider the mass combination $H + HL \rightarrow HH + L$ on a repulsive surface, the reverse situation holds. The parameter β is now very much less than unity and it is the entrance channel that is long and narrow whereas the exit channel is broad and flat, as shown in figure 6.9. Thus the approach of H to HL will be slow and when the particle reaches the head of the entrance valley there are many possible diagonal paths down the exit valley. Thus mixed energy release can occur with substantial vibrational excitation. Mixed energy release is favoured for surfaces for which the scaling factor β is small. It is also favoured if the skewing angle ϕ is large, as is the case when $m_A \approx m_C \gg m_B$. If the axes are strongly skewed, the particle can bounce off the repulsive wall of the entrance valley, enter the exit valley from the side and hence acquire substantial vibrational excitation.

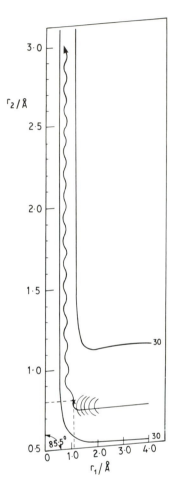

Figure 6.8. Collinear trajectory for mass combination L + HH on surface I. (Adapted from B. A. Hodgson and J. C. Polanyi, *J. Chem. Phys.*, **55**, 4745 (1971).)

The angular distribution of the products is also dependent on the attractive nature of the potential surface. We define the forward direction as being in the direction of the relative velocity of A with respect to BC. As the amount of attractive energy release increases, there is an increased tendency for forward scattering to occur. This is illustrated in figure 6.10 for all of the various mass combinations involving light (L) and heavy (H) atoms on the three surfaces of figure 6.6. The mass combination L + HH is also anomalous with regard to angular distribution. Even for very attractive surfaces, the scattering of the product LH is predominantly backward.

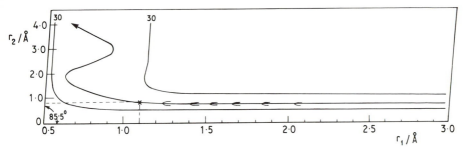

Figure 6.9. Collinear trajectory for mass combination H + HL on surface I. (Adapted from Hodgson and Polanyi 1971.)

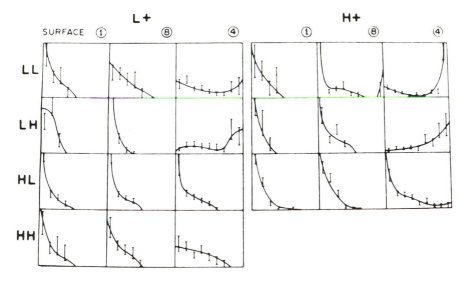

Figure 6.10. Angular distribution of products for various mass combinations on surfaces (1), (8) and (4) of figure 6.6. Left-hand side of each diagram corresponds to backward scattering. (Reproduced from Polanyi and Schreiber 1974.)

Collinear calculations ignore rotational motion and cannot predict the occurrence of rotational excitation. This will be likely to occur for large impact parameter collisions on attractive surfaces. The initial angular momentum is large and this is converted into rotational excitation of the product. In the case of repulsive surfaces, rotational excitation can occur by the release of BC repulsion for bent ABC configurations.

6.3.3. The Effects of Barrier Location

The location of the barrier in the potential energy surface has a pronounced effect of the dynamics of a reaction. We discuss this from the point of view of a thermoneutral reaction and consider the two potential energy surfaces illustrated in figure 6.11.

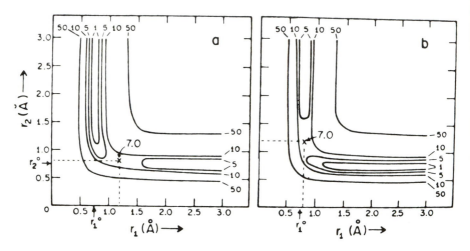

Figure 6.11. Thermoneutral collinear potential surfaces with barrier (*a*) in entrance valley; (*b*) in exit valley. (Reproduced from Polanyi and Wong 1969.)

Both have a barrier of approximate height $29 \, \text{kJ} \, \text{mol}^{-1}$ but for surface I the crest of the barrier is in the entrance valley at $r_1^{\ddagger} = 1.2 \, \text{Å}$ and $r_2^{\ddagger} = 0.8 \, \text{Å}$, whereas for surface II the crest of the barrier is in the exit valley at $r_1^{\ddagger} = 0.8 \, \text{Å}$, and $r_2^{\ddagger} = 1.2 \, \text{Å}$. Energy profiles for the two surfaces are given in figure 6.12 and show that although the crest may be in the entrance valley or in the exit valley, the barrier extends into both valleys.

Polanyi and Wong (1969) considered the mass combinations L + LL in a set of three-dimensional trajectory calculations. Comparable results will be obtained for other mass combinations. The location of the crest of the barrier has two effects. The differing effects of initial relative translational energy and of vibrational energy were investigated. In order for reaction to occur it is clearly necessary for the system to have an energy in excess of the height of the barrier, but the reaction cross-section is very dependent on the nature of the initial energy. For surface I it was found that the reaction cross-section has a non-zero value for translational energies only slightly in excess of the barrier height and the cross-section rises rapidly as the kinetic energy increases. By contrast initial energy in the form of vibrational energy

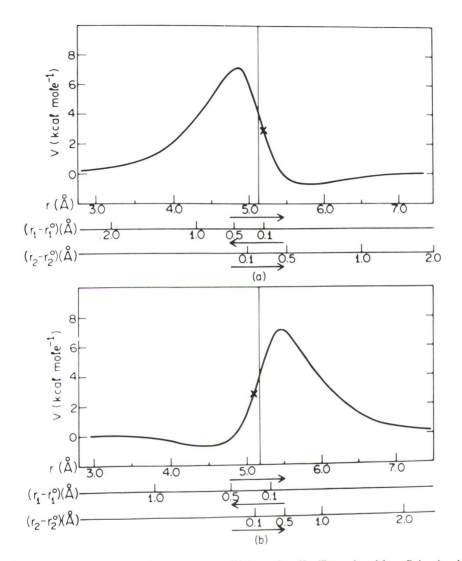

Figure 6.12. Energy profiles (*a*) for surface I; (*b*) for surface II. (Reproduced from Polanyi and Schreiber 1974.)

is ineffective in promoting reaction and the cross-section is zero even when the vibrational energy is equal to twice the barrier height. The opposite situation was found for surface II, where the reaction occurs if the vibrational energy is slightly in excess of the barrier height and the cross-section rises rapidly with increasing vibrational excitation. The cross-section is zero for translational energies up to twice the barrier height.

This difference can be interpreted in terms of trajectories on a collinear surface. In order for reaction to occur, motion in the entrance valley must be converted into motion in the exit valley. If the barrier is in the entrance valley, translational energy will be effective in surmounting the barrier, but if the initial energy is in the form of vibrational energy, the trajectory will be reflected by the barrier and reaction will not occur. On surface II, where the barrier is in the exit valley, if the particle has translational energy the trajectory will be reflected by the repulsive wall at the head of the entrance valley and the barrier will not be surmounted. However, vibrational motion will be effective in enabling the trajectory to turn the corner and surmount the barrier.

The location of the barrier also has an effect on the conversion of the initial energy of the reactants into the final energy of the products. On surface I, where initial translational energy is required to surmount the barrier, an amount of initial translational energy close to threshold is converted into vibrational energy of the product molecule. This can be understood in terms of the initial translational energy being in the direction of r_1 which corresponds to the vibrational motion of the product molecule AB. Conversely on surface II, for which initial vibrational energy is a prerequisite for reaction to occur, the initial motion is in the direction of r_2 and an amount of initial vibrational energy (close to threshold) of the reactant molecule is converted into relative translational energy of the products. This is illustrated for collinear trajectories with the mass combinations L + HH and H + HL on surfaces I and II as shown in figures 6.8 and 6.9. As discussed above, these two mass combinations represent two extremes of scaling with β being very large for L + HH and very small for H + HL.

If the initial energies are substantially above threshold, the excess energy is not converted from translational energy to vibrational energy or vice versa. Excess translational energy in the reactants appears as translational energy of the products and similarly for excess vibrational energy.

The conclusions derived from consideration of thermoneutral potential surfaces also apply to exothermic and endothermic reactions. For exothermic processes, the crest of the barrier generally lies in the entrance valley irrespective of the attractive or repulsive character of the surface. Thus for an exothermic reaction, translational energy is effective in promoting reaction. Conversely, for endothermic reactions the crest of the barrier will be in the exit valley and vibrational excitation will be required in order to promote reaction. This has been demonstrated experimentally for the reaction

$$HCl + K \rightarrow H + KCl \qquad (6.88)$$

for which the cross-section for $HCl(v = 1)$ is two orders of magnitude larger than that for $HCl(v = 0)$. There have been several trajectory calculations on endothermic surfaces which confirm that vibrational energy is effective in promoting reaction.

6.3.4. *Secondary Encounters and Complex Trajectories*

The discussion in the previous section has assumed that trajectories do not involve secondary encounters in which the product molecule AB undergoes a subsequent interaction with atom C. Because it is not possible to discuss secondary encounters in terms of collinear trajectories, it is not easy to make generalizations about the outcome of secondary encounters, but in general they lead to broadening of the energy and angular distributions of the products. Secondary encounters are likely to occur on very attractive surfaces where there is very little repulsive energy to separate the products. Polanyi (see Polanyi and Schreiber 1974) distinguishes two such types of encounter. In the reaction

$$A + BC \rightarrow AB + C \tag{6.89}$$

a repulsive secondary encounter may occur between atom C and atom A as the product molecule AB rotates in such a way as to reverse the direction of rotation of AB. This is termed 'clouting'. The atom C is scattered forward and the molecule AB backwards. Such secondary encounters lead to very little rotational excitation of the product molecule. A different type of secondary encounter occurs in the reaction of A with BC when it appears initially that the product AB is going to be formed but an attractive interaction between A and C ('clutching') reverses the direction of A leading to the formation of the product AC. In this case the product AC is strongly scattered forward. This latter type of trajectory is sometimes termed a migratory encounter because A initially strikes B but a secondary encounter of A with C leads to the formation of the product AC.

Secondary encounters can also occur, but less commonly, on repulsive surfaces with mass combinations $m_A \approx m_C \gg m_B$ for which a very high proportion of the repulsion is converted into internal excitation.

Reaction on potential surfaces containing hollows often involves an indirect mechanism. Trajectory calculations have been done on surfaces with a hollow of depth $29 \, \text{kJ mol}^{-1}$ in either the entrance or the exit valley. These surfaces are analogous to surfaces I and II discussed in section 6.3.3. The trajectories were not so indirect that one could say that there was a collision complex with a lifetime larger than one rotational period. The angular distributions were found to be broad and this is a consequence of the indirect nature of the trajectories. If the hollow is in the entrance valley, vibrational excitation is required for reaction to take place, whereas translational energy is effective for a hollow in the exit valley. This is the converse of the situation for surfaces with a potential barrier.

If a collision complex is formed which has a lifetime of several rotational periods, the angular distribution is found to be symmetric with respect to the scattering angle of 90°. Such behaviour is often observed in trajectory calculations in which there is a deep potential well. However a deep well in the potential surface does not necessarily result in the formation of a collision complex.

6.3.5. *Results of Trajectory Calculations for the Reaction of F with H_2*

There is much detailed experimental data for the reaction

$$F + H_2 \rightarrow HF + H \qquad (6.90)$$

and its isotopic variants which have been investigated by infra-red chemi-luminescence, the crossed molecular beam method and by chemical laser experiments. A comprehensive review of this reaction has been given by Anderson (1980). The major part of the exothermicity of the reaction goes into vibrational excitation of the product, and from the infra-red chemi-luminescence and chemical laser experiments it is possible to determine the vibrational and rotational energy distributions in the product molecule. This abundance of experimental data has stimulated many quasi-classical trajec-tory studies of this reaction in order to see how successful the method is in interpreting the existing data and predicting hitherto unobserved phenomena. This work has been reviewed in detail by Muckerman (1981) so here we attempt to summarize the main conclusions of these studies. Polanyi and Schreiber (1977) also present a comprehensive discussion of this system.

The calculations have attempted to provide several sorts of information such as the thermal rate constant, detailed rate constants for various internal states of H_2, excitation functions, detailed rate constants for the formation of HF in various final states, rotational, vibrational and angular distributions of the products and the vibrational–rotational energy distribution in the pro-ducts.

Agreement with experiment is rather poor for the calculation of the thermal rate constant for the reaction with H_2. The values obtained in quasi-classical trajectory calculations at $300\,K$ are in the range $0.5–7\times10^{-12}\,cm^3\,molecule^{-1}\,s^{-1}$ which is well below the range of values for reliable experimental determinations ($19–37\times10^{-12}\,cm^3\,molecule^{-1}\,s^{-1}$). It is unlikely that such a large discrepancy can be accounted for by quantum mechanical tunnelling.

The barrier for the $F + H_2$ surface (see figure 3.1 for a typical surface) lies in the entrance valley and it would therefore be expected from the discus-sion in section 6.3.3 that translational energy would be more effective than vibrational energy in surmounting the barrier. This is confirmed in calcula-tions which show that translational energy is about six times more effective than vibrational energy. The total reactive cross-section is predicted to rise smoothly from a threshold at an energy approximately equal to the barrier height to a value of about $4\,\text{Å}^2$ at an initial relative energy E_{rel} of

$25 \, \text{kJ mol}^{-1}$. The effect of rotational energy has also been investigated. For a fixed value of E_{rel}, the cross-section increases in going from $J = 0$ to $J = 1$ but decreases as J is increased further. The state-selected rate constants $k_r^{v,J}(T)$ for H_2 in the initial vibration–rotation state (v, J) also show a strong dependence on J with the Arrhenius activation energy increasing as J increases. It is not at present possible to select experimentally the rotational state of H_2 so there are no experimental data with which these results can be compared. The opacity function falls off gradually as the impact parameter b increases and the largest value for which reaction occurs is $b_{\text{max}} = 1.77 \, \text{Å}$. This behaviour is typical for low energy reactions on potential surfaces having a potential barrier. Comparable results are obtained for the reaction of F with D_2 and the ratio of the thermal rate constants for the H_2 and D_2 reactions is in accord with experiment.

In the reaction of F with HD, there are two possible reaction channels

$$F + HD \rightarrow HF + D \tag{6.91}$$

or

$$F + HD \rightarrow DF + H. \tag{6.92}$$

Trajectory calculations for low values of E_{rel} show that the ratio of the reactive cross-sections $\sigma_R^{HF}/\sigma_R^{DF}$ for reactions 6.91 and 6.92 depends strongly on the value of J. If there is no rotational energy, fluorine will react preferentially at the D end of HD. Examination of the skewed potential energy surfaces for collinear FHD and FDH reveals that for FDH trajectories are less likely to be reflected back into the entrance channel by the repulsive wall of the product valley. For $J = 0$, the thresholds for both reactions are comparable but for $J \geqslant 2$ the DF channel has a higher threshold and for $J = 4$ the cross-section for the formation of DF is lower than that for HF. This is interpreted in terms of it not being possible for the F to get close to the D before it is intercepted by the H atom. For the two channels the rate constants for state-selected reactants show different behaviour with respect to J but on thermal averaging the two values are very similar.

We turn now to the consideration of product distributions. For E_{rel} less than $0.5 \, \text{eV}$ $(48 \, \text{kJ mol}^{-1})$, trajectory calculations give angular distributions in which the product HF or DF is backward scattered but for energies above $5 \, \text{eV}$ $(480 \, \text{kJ mol}^{-1})$ forward scattering is obtained. For a value of $E_{\text{rel}} = 12.5 \, \text{kJ mol}^{-1}$ the angular distributions for the reactions $F + H_2$ and $F + D_2$ are virtually the same. The distributions for HF and DF from the reaction of F with HD are also very similar to each other. The results for $F + D_2$ at low energy are consistent with experiment.

One of the prominent features of the $F + H_2$ reaction is that most of the available energy is converted into vibrational energy of HF. An inverted vibrational distribution is produced in which the maximum population is in level $v'_{\text{max}} - 1$, where v'_{max} is the highest level that can be populated. For HF the $v' = 2$ level is preferentially populated and this is correctly reproduced by all trajectory calculations, on a variety of potential energy surfaces, for this reaction. With one exception, the calculations indicate that about 71% of the available energy is channelled into HF vibration, in agreement with experiment. However it is difficult to say which potential surface is to be preferred. Although the calculations give a maximum population for $v' = 2$, the continuous classical distribution of vibrational energies has a peak between $v' = 2$ and $v' = 3$ (for H_2 initially in the state $v = 0$, $J = 1$ and $E_{rel} = 12.5 \, kJ \, mol^{-1}$). This is completely different from the experimental distribution and from the results of quantum mechanical scattering calculations. For the reaction with D_2, most of the calculations result in a maximum population for $v' = 3$, in accord with experiment, but the calculated vibrational distributions are too sharply peaked. In the case of the reaction with HD, it is found experimentally that a smaller fraction of the total energy goes into HF vibration than into DF vibration. This has been interpreted in terms of special dynamical requirements for the formation of HF with $v' = 3$. Trajectory calculations give the opposite result that a larger fraction of the energy appears as HF vibration.

Rotational state distributions for the products can be obtained from infra-red chemiluminescence experiments. The fraction of the available energy converted into rotational energy decreases along the series $FH + D$, $FH + H$, $FD + D$, $FD + H$ and trajectory calculations are able to reproduce this trend.

So far we have discussed the distribution of just one variable of the product molecule. From infra-red chemiluminescence studies one can obtain both the vibrational and rotational energy distributions of the product molecule and from molecular beam experiments one obtains the velocity distribution of the product molecule as a function of the scattering angle. Analysis of the results of trajectory calculations in terms of two variables of the product is less satisfactory than analysis in terms of one variable because of statistical noise.

Calculations for $F + H_2$ by Muckerman (1981) of the rotational distribution as a function of the vibrational quantum number gave average rotational excitations $\langle J' \rangle_{v'}$ which are larger than the experimental values and which also increase faster as v' decreases. Polanyi and Schreiber (1977) found that for $v' = 1$ and $v' = 3$, the calculated values of $\langle J' \rangle_{v'}$ were larger than the experimental values but for $v' = 2$ the reverse was found. Molecular beam studies of the reaction of F with D_2 show well resolved vibrational structure in the contour diagrams representing the velocity of the product as a function of scattering angle. Trajectory calculations are unable to reproduce this result and either fail to give appreciably different angular distributions for different values of v', the vibrational state of the product, or predict a small shift in the mean scattering angle as v' increases.

Thus quasi-classical trajectory calculations do reproduce the main features of the $F + H_2$ reaction such as the vibrational distribution of the products, the fraction of energy being converted into vibrational energy of the product and the backward scattering observed in molecular beam experiments. However they are less successful in yielding the detailed vibration–rotation distribution and the angular–velocity distribution.

6.4. Non-Adiabatic Reactions

The quasi-classical trajectory method outlined in the previous sections assumes that the motion occurs on a single potential energy surface and is not directly applicable to reactions involving a transition from one electronic state to another. Such non-adiabatic transitions occur in charge transfer processes and are common in ion–molecule reactions where potential surfaces are often in close proximity. The quasi-classical trajectory method has been extended to deal with such processes and in this section we discuss the surface hopping trajectory method developed by Tully and Preston (1971). It is assumed that if the interaction between the two surfaces is small then motion will occur on one surface only and the trajectories can be calculated by the methods outlined above. However if the two surfaces are in close proximity, the Born–Oppenheimer separation is no longer valid and there is coupling between the surfaces. There is then a finite probability of a transition taking place from one surface to the other. This probability is estimated and the calculation proceeds by following trajectories on both surfaces. In the final averaging process each trajectory is weighted according to its probability. We discuss the method in more detail below and give an example of its application.

Since the Born–Oppenheimer separation is not valid, we have to return to the coupled equations 1.15

$$[T_{\text{nuc}} + U_i(\boldsymbol{R}) + B_{ii}(\boldsymbol{R}) - E]\chi_i(\boldsymbol{R}) = -\sum_{j \neq i} c_{ij}(\boldsymbol{R}, \boldsymbol{P})\chi_j(\boldsymbol{R}) \tag{6.93}$$

for the nuclear wavefunction $\chi_i(\boldsymbol{R})$. T_{nuc} is the nuclear kinetic energy operator, $U_i(\boldsymbol{R})$ is the potential energy function for the electronic state i and c_{ij} represents the non-adiabatic coupling between electronic states i and j. Each of the coefficients $c_{ij}(\boldsymbol{R}, \boldsymbol{P})$ is the sum of two terms

$$c_{ij}(\boldsymbol{R}, \boldsymbol{P}) = \sum_k \frac{1}{M_k}(A_{ij}^{(k)} \cdot \boldsymbol{P}_k + B_{ij}^{(k)}) \tag{6.94}$$

which are expressed in terms of integrals with respect to the electronic wavefunctions $\Phi_i(\boldsymbol{R}, \boldsymbol{r})$

$$A_{ij}^{(k)}(\mathbf{R}) = \int \Phi_i^*(\mathbf{R}, r)\left(-\frac{ih}{2\pi}\nabla_k\right)\Phi_j(\mathbf{R}, r)\ d\tau \qquad (6.95)$$

and

$$B_{ij}^{(k)}(\mathbf{R}) = \int \Phi_i^*(\mathbf{R}, r)\left(-\frac{h^2}{8\pi^2}\nabla_k^2\right)\Phi_j(\mathbf{R}, r)\ d\tau. \qquad (6.96)$$

Coupling between two surfaces may also occur by spin–orbit coupling and if this effect is important the appropriate off-diagonal spin–orbit interaction integrals

$$\int \Phi_i^* H_{SO}\Phi_j\ d\tau$$

must be included in the right-hand side of equation 6.93 with the corresponding diagonal term being added to the left-hand side. The general formulation outlined above leads to an infinite set of coupled equations 6.93 for the nuclear wavefunction and the problem of non-adiabatic processes is tractable only if it is possible to choose a set of electronic functions $\Phi_i(\mathbf{R}, r)$ in such a way that only a few of the coupling terms c_{ij} are non-zero.

If potential surfaces are well separated, then the terms c_{ij} $(i \neq j)$ are negligible and the methods discussed in sections 6.2 and 6.3 are applicable. Non-adiabatic coupling is expected to be important when potential surfaces are close to each other and Massey proposed that the system will behave adiabatically if

$$\frac{h}{2\pi}\frac{\mathbf{v}\cdot\mathbf{d}_{ij}^{(k)}}{|E_i - E_j|} \ll 1, \qquad (6.97)$$

where $\mathbf{v}$ is the classical nuclear velocity, $\mathbf{d}_{ij}^{(k)}$ is the integral

$$\int \Phi_i\nabla_k\Phi_j\ d\tau$$

and E_i and E_j are the energies of states i and j.

The dominant term in the non-adiabatic coupling is usually $A_{ij}^{(k)}\cdot\mathbf{P}_k/M_k$ and $B_{ij}^{(k)}$ is usually very much smaller. There are two types of contribution to the term $A_{ij}^{(k)}$ arising from rotational and radial coupling. For non-adiabatic transitions it is the radial coupling that is of primary importance. In any system where non-adiabatic processes are expected to be important, it is necessary to identify the regions where the coupling is strong by calculating the vector $\mathbf{d}_{ij}^{(k)}$ which can be expressed in terms of integrals of the type

$$\int \Phi_i(\mathbf{R}, r)\frac{\partial\Phi_j(\mathbf{R}, r)}{\partial R}\ d\mathbf{r}$$

which can be evaluated by finite difference methods (see Buenker *et al.* 1982). A graphical representation of $\mathbf{d}_{ij}^{(k)}$ can be obtained by plotting one component in the form of a contour diagram for a particular geometrical configuration. The regions where non-adiabatic interactions are important appear as seams on the contour diagram. An example of such a diagram is given in figure 6.13 for collinear geometries of the H_3^+ system, where non-adiabatic effects arise from the avoided intersection between potential surfaces for the reaction

$$H^+ + H_2 \rightarrow H_2 + H^+ \qquad (6.98)$$

and

$$H^+ + H_2 \rightarrow H_2^+ + H. \qquad (6.99)$$

The origin of the avoided intersection can be seen by plotting cuts through the collinear potential surfaces for the interaction of $H + H_2^+$ and of $H^+ + H_2$ as functions of R_1 (see figure 6.14) with R_2 infinite. The two potential curves will cross in the region of $R_1 = 2.5\,a_0$ $(1\,a_0 = 5.2917706 \times 10^{-11}\,\text{m})$ but the crossing is avoided. The strong interaction persists as R_2 is decreased to about $6\,a_0$. The seam of the avoided crossing is clearly seen in figure 6.13. As we discussed in section 3.3, two potential surfaces depending on m variables will have a surface of intersection of dimension $m-2$. The quantity d_{ij} is a vector and thus some directions of motion will be more effective than others in promoting non-adiabatic transitions. In the case of the H_3^+ system it is vibrational energy that is effective.

We turn now to consideration of the dynamical aspects of non-adiabatic transitions and start with an outline of the classical path approach, in which electronic motion is considered from the point of view of quantum mechanics and nuclear motion is treated classically. This approach is clearly valid only in situations where the classical trajectory method would be appropriate in the absence of electronic transitions. This will be true in regions well away from the avoided crossing. Also it can be shown that the classical path approach is valid if the difference in energy between the two interaction potential surfaces is small compared with the kinetic energy T of the nuclei, i.e.

$$\frac{E_2 - E_1}{T} \ll 1, \qquad (6.100)$$

and if the off-diagonal interaction term is small near the classical turning point or if the forces have the same sign near the turning point.

The nuclear motion is assumed to be described by a classical trajectory $R(t)$ and the time-dependent electronic wavefunction $\Phi(r,t)$ satisfies the time-dependent Schrödinger equation

$$H_{\text{el}}(R,r)\Phi(r,t) = \frac{ih}{2\pi}\frac{\partial}{\partial t}\Phi(r,t), \qquad (6.101)$$

where $H_{\text{el}}(R,r)$ is the electronic Hamiltonian as defined in equation 1.3. The wavefunction $\Phi(r,t)$ can be expressed in terms of the electronic

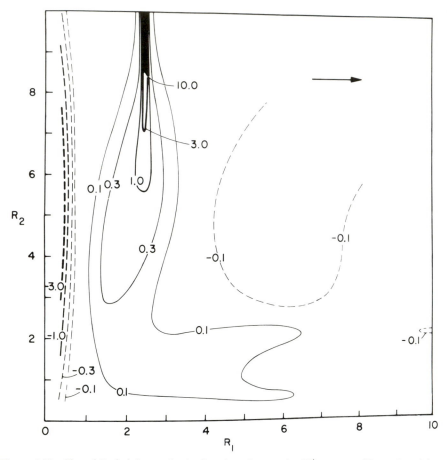

Figure 6.13. Plot of $D_{12}(\rightarrow)$ for motion in direction of arrow for H_3^+ system. (Reproduced from Tully and Preston 1971.)

wavefunctions $\Phi_i(\boldsymbol{R}, \boldsymbol{r})$ for the stationary states using the standard methods of time-dependent perturbation theory thus:

$$\Phi(\boldsymbol{r}, t) = \sum_j a_j(t)\Phi_j(\boldsymbol{R}, \boldsymbol{r})\exp\left(-\frac{2\pi\mathrm{i}}{h}\int_0^t W_{jj}(\boldsymbol{R})\,\mathrm{d}t\right), \qquad (6.102)$$

where $a_j(t)$ are time-dependent coefficients and

$$W_{ij} = \int \Phi_i(\boldsymbol{R}, \boldsymbol{r})H_{\mathrm{el}}\Phi_j(\boldsymbol{R}, \boldsymbol{r})\,\mathrm{d}\tau. \qquad (6.103)$$

Substitution of the expansion of equation 6.102 into equation 6.101,

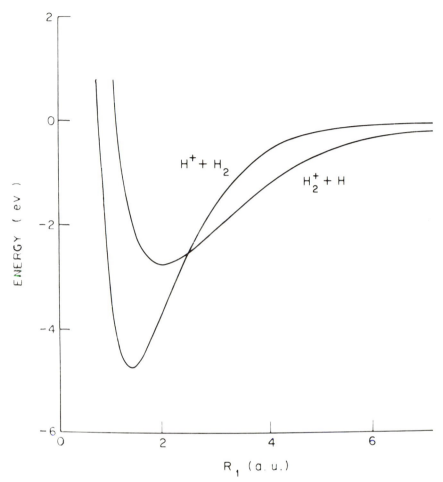

Figure 6.14. Cuts through the lowest singlet potential surfaces of H_3^+ for collinear geometries as a function of R_1, the distance between atoms 1 and 2, with R_2, the distance between atoms 2 and 3, infinite. These curves are equivalent to potential curves for H_2 and H_2^+ drawn relative to a common dissociation limit. (Reproduced from Tully and Preston 1971.)

multiplication on the left by $\Phi_i(\mathbf{R}, \mathbf{r})$ and integration with respect to the electronic coordinates yields the following equation for the time derivative $\dot{a}_i$ of a_i:

$$\frac{ih}{2\pi} \dot{a}_i = \sum_{j \neq i} a_j \left(W_{ij} - \frac{ih}{2\pi} \int \Phi_i(\mathbf{R}, \mathbf{r}) \frac{\partial \Phi_j(\mathbf{R}, \mathbf{r})}{\partial t} \, d\mathbf{r} \right)$$

$$\times \exp\left(-\frac{2\pi i}{h} \int_0^t (W_{jj} - W_{ii}) \, dt \right). \qquad (6.104)$$

The integral

$$\int \Phi_i(\boldsymbol{R},r)\frac{\partial \Phi_j(\boldsymbol{R},r)}{\partial t}\,\mathrm{d}r.sp.25$$

can be rewritten using the chain rule as

$$\dot{\boldsymbol{R}}\cdot\int \Phi_i(\boldsymbol{R},r)\nabla_R\Phi_j(\boldsymbol{R},r)\,\mathrm{d}r \qquad (6.105)$$

and thus

$$\frac{ih}{2\pi}\dot{a}_i = \sum_{j\neq i}a_j\left(W_{ij}-\frac{ih}{2\pi}\dot{\boldsymbol{R}}\cdot\int \Phi_i(\boldsymbol{R},r)\nabla_R\Phi_j(\boldsymbol{R},r)\,\mathrm{d}r\right)$$

$$\times\exp\left(-\frac{ih}{2\pi}\int_0^t (W_{jj}-W_{ii})\,\mathrm{d}t\right). \qquad (6.106)$$

From this set of coupled equations one can calculate the probability $|a_i|^2$ that the system is in state i after time t for any trajectory $\boldsymbol{R}(t)$. If the electronic wavefunctions are the adiabatic functions, the off-diagonal terms W_{ij} are zero and the coupling is given by the term

$$\int \Phi_i(\boldsymbol{R},r)\nabla_R\Phi_j(\boldsymbol{R},r)\,\mathrm{d}r$$

as in equation 6.95. If there are just two interacting adiabatic electronic states $\Phi_1(\boldsymbol{R},r)$ and $\Phi_2(\boldsymbol{R},r)$ then

$$\dot{a}_1 = -a_2\dot{\boldsymbol{R}}\cdot\int \Phi_1(\boldsymbol{R},r)\nabla_R\Phi_2(\boldsymbol{R},r)\,\mathrm{d}r\,\exp\left(-\frac{ih}{2\pi}\int_0^t (W_{22}-W_{11})\,\mathrm{d}t\right) \quad (6.107)$$

and similarly for $\dot{a}_2$.

Although it is now possible to obtain exact numerical solutions to the set of equations for the one-dimensional problem, the Landau–Zener–Stueckelberg approximation is frequently used for two-state systems for which there is only one internal coordinate R. The velocity $\dot{R}$ is assumed to be constant in the interaction region and the matrix $\boldsymbol{W}$ (for the integrals of equation 6.103 in terms of diabatic wavefunctions $\Phi_i(\boldsymbol{R},r)$) is of the form

$$\boldsymbol{W} = \begin{pmatrix} W_0+\tfrac{1}{2}B(R-R_0) & A \\ A & W_0-\tfrac{1}{2}B(R-R_0) \end{pmatrix}, \qquad (6.108)$$

where A and B are constants and R_0 is the value of R for which the diabatic curves intersect. The probability P_{12} that a system initially in state 1 will undergo a transition to state 2 is given by solving equation 6.104 to give

$$P_{12} = \exp\left(-\frac{4\pi^2 A^2}{B\dot{R}h}\right). \qquad (6.109)$$

The approximation will be poor at low collision energies and when the avoided intersection is near a classical turning point, because the velocity will not be constant under such circumstances. Similarly at high collision energies one cannot assume that the off-diagonal coupling A is constant.

In the surface hopping trajectory method of Tully and Preston (1971) it is necessary to specify two or more N-dimensional potential energy surfaces (which may be adiabatic or diabatic surfaces) and the off-diagonal interaction between the surfaces. It is also necessary to define the region for which non-adiabatic transitions can take place in the form of a relation of the type

$$S(\mathbf{R}) = 0. \tag{6.110}$$

$S(\mathbf{R})$ defines an $(N-1)$-dimensional avoided-crossing hypersurface. A trajectory is started on one surface with the initial conditions being selected by the methods discussed in section 6.2.2 and integration is continued until the condition $S(\mathbf{R}) = 0$ is satisfied. At this point the trajectory is split into n branches where n is the number of potential energy surfaces that strongly interact in that region. The probability $P_i(\mathbf{R}, \dot{\mathbf{R}})$ of the trajectory undergoing a transition to surface i is obtained by numerical integration of the coupled equations 6.104 for the amplitudes a_i through the strong coupling region and computing the probability from

$$P_i(\mathbf{R}, \dot{\mathbf{R}}) = |a_i|^2. \tag{6.111}$$

The calculation then proceeds by the integration of each of the trajectories on the interacting surfaces. If any of them reach the interaction region again, the surface hopping procedure is repeated. In the final averaging process to calculate cross-sections or detailed rate constants each of the branched trajectories is weighted by the appropriate $P_i(\mathbf{R}, \dot{\mathbf{R}})$ factor. In calculations for the $H^+ + H_2$ system, Tully and Preston (1971) found that the transition probabilities were given satisfactorily by the Landau–Zener–Stueckelberg approximation. If adiabatic potential surfaces are used, there is an additional complication arising from the fact that at the avoided crossing the surfaces are not degenerate. This is taken into account by adjusting the velocity along the non-adiabatic coupling vector to conserve energy.

Comparison of the results of surface hopping trajectory calculations of Krenos *et al.* (1974) for the reaction

$$D^+ + HD \rightarrow H^+ + D_2 \tag{6.112}$$

using DIM potential surfaces, shows that the method is capable of giving good agreement with experimental data.

For a full discussion of the method the reader is referred to the article by Tully (1976). Non-adiabatic processes are discussed in detail by Nikitin (1970) and Child (1979).

164   *Potential Energy Surfaces*

Suggestions for Further Reading

R. B. Bernstein (1982) *Chemical Dynamics via Molecular Beam and Laser Techniques,* Clarendon Press, Oxford.

R. D. Levine and R. B. Bernstein (1974) *Molecular Reaction Dynamics,* Clarendon Press, Oxford.

J. C Polanyi (1972) *Acc. Chem. Res.,* **5**, 161.

I. W. M. Smith (1980) *Kinetics and Dynamics of Elementary Gas Reactions,* Butterworths, London.

D. G. Truhlar (ed.) (1981) *Potential Energy Surfaces and Dynamics Calculations,* Plenum Press, New York.

References

J. B. Anderson (1980) *Adv. Chem. Phys.,* **41**, 229.

R. B. Bernstein (1979) *Atom–Molecule Collision Theory—A Guide for the Experimentalist,* Plenum, New York.

R. J. Buenker, G. Hirsch, S. D. Peyerimhoff, P. J. Bruna, J. Römelt and M. Bettendorf (1982) in *Studies in Physics and Theoretical Chemistry 21, Current Aspects of Quantum Chemistry 1981,* ed. R. Carbó, p. 81, Elsevier, Amsterdam.

D. L. Bunker (1971) *Methods in Computational Physics,* **10**, 287.

M. S. Child (1974) *Molecular Collision Theory,* Academic Press, London.

M. S. Child (1976) in *Dynamics of Molecular Collisions,* part B, ed. W. H. Miller, p. 171, Plenum Press, New York.

M. S. Child (1979) in Bernstein (1979), p. 427.

J. N. L Connor (1979) *Comp. Phys. Comm.,* **17**, 117.

H. Goldstein (1950) *Classical Mechanics,* Addison-Wesley, Reading, MA.

M. Karplus, R. N. Porter and R. D. Sharma (1965), *J. Chem. Phys.,* **43**, 3259.

J. R. Krenos, R. K. Preston, R. Wolfgang and J. C. Tully (1974) *J. Chem. Phys.,* **60**, 1634.

P. J. Kuntz (1976) in *Dynamics of Molecular Collisions,* part B, ed. W. H. Miller, p. 53, Plenum Press, New York.

P. J. Kuntz, E. M. Nemeth, J. C. Polanyi, S. D. Rosner and C. E. Young (1966) *J. Chem. Phys.,* **44**, 1168.

A. Kupperman (1981) in *Theoretical Chemistry* Vol. 6A, ed. D. Henderson, p. 80, Academic Press, New York.

W. H. Miller (1976) *Dynamics of Molecular Collisions,* Plenum Press, New York.

J. T. Muckerman (1981) in *Theoretical Chemistry* Vol. 6A, ed. D. Henderson, p. 1, Academic Press, New York.

E. E. Nikitin (1970) *Adv. Quantum Chem,* **5**, 135.

J. C. Polanyi and J. L. Schreiber (1974) in *Physical Chemistry: An Advanced Treatise* Vol. 6A, ed. W. Jost, p. 383, Academic Press, New York.

J. C. Polanyi and J. L Schreiber (1977) *Farad. Disc. Chem. Soc.,* **62**, 267.

J. C. Polanyi and W. H. Wong (1969) *J. Chem. Phys.,* **51**, 1439.

R. N. Porter and L. M. Raff (1976) in Miller (1976), part B, p. 1.

D. G. Truhlar and J. T. Muckerman (1979) in Bernstein (1979), p. 505.

J. C. Tully (1976) in Miller (1976), part B, p. 217.

J. C. Tully and R. K. Preston (1971) *J. Chem. Phys.,* **55**, 562.

CHAPTER 7

experimental studies of
reaction dynamics

7.1. Molecular Beam Scattering

When an atom or molecule collides with another atom or molecule, scattering occurs in which the direction of the relative velocity vector changes. There are three different possibilities. In elastic scattering there is no chemical reaction or change in the internal states of the species undergoing collision and the magnitude of the relative velocity remains unchanged. Second, if there is a change in the internal state of one of the collision partners, but no chemical reaction, the scattering is inelastic. The third case is that of reactive scattering in which a chemical reaction occurs. The magnitude of the relative velocity does change in inelastic and reactive scattering. The dynamics of the scattering are clearly dependent on the potential energy surface for the system and from the investigation of the dynamics, deductions can be made regarding the nature of the surface. In the case of elastic scattering in atom–atom collisions, the potential energy depends only on the interatomic distance and in cases where the data are of sufficient accuracy it is possible to obtain the interatomic potential by inversion of the scattering data. This is not possible for atom–molecule potentials and for such systems it is necessary to compare the experimental data with the results of dynamical calculations on model potential surfaces. In some cases it is possible to vary the parameters of the model potential to achieve agreement with experiment. In this chapter we will start by outlining the experimental techniques used and the transformation of the data from the laboratory coordinate system to the centre-of-mass system used in theoretical work. In subsequent sections we will discuss the determination of interatomic potentials from elastic scattering and the dynamics of reactive systems.

7.1.1. Molecular Beam Techniques

The ultimate aim of molecular beam studies is to observe the collision between two state-selected species with a well defined initial relative velocity under single collision conditions. The deflection and the velocity distribution of the scattered atom or molecule are measured and ideally one would like to determine the vibration–rotation states of scattered molecules. If this is done under conditions such that the species being observed does not undergo further collisions before observation, then deductions can be made about the dynamics of the collision. Unfortunately it is not yet possible to perform experiments which give the level of detail specified above.

In order to achieve single collision conditions it is necessary to work at pressures of the order of 10^{-6} to 10^{-7} torr (1 torr $= 1.333\,22 \times 10^2\,\mathrm{N\,m^{-2}}$). A schematic diagram of the molecular beam experiment was given in figure 6.1. Two well collimated molecular beams intersect at right angles in a large chamber in which the background pressure is less than 10^{-6} torr. More detailed information can be obtained about the scattering process if the beams have well defined velocities. Collisions between the species in the two beams occur in the small volume in which the beams cross. In the simplest experiments the intensity of the scattered species is measured as a function of scattering angle by a suitable detector which can be rotated, inside the vacuum chamber, in the plane defined by the two reactant beams. It is now routine to measure the velocity (or energy) distribution of the scattered species and in some cases it is possible to obtain information about its internal energy. We will discuss briefly molecular beam sources, detectors, velocity and state selection and the measurement of the velocity distribution and internal state of the scattered species. For more comprehensive discussions the reader should consult Fluendy and Lawley (1973), Toennies (1974), Gentry (1979), Grice (1981a) or Bernstein (1982). In some experiments a beam of one species passes through a collision cell containing the scattering gas. Such beam–gas experiments will not give the detailed angular and velocity distributions that can be obtained from crossed-beam experiments but the observation of chemiluminescence or laser-induced fluorescence from product molecules gives a detailed picture of the internal energy distribution in the product molecule.

7.1.1.1. Molecular Beam Sources and State Selection

In the earlier work with neutral species, effusive molecular beam sources were used. The condition for effusive flow is that the diameter of the orifice or the width of the slit in the source should be larger than the mean free path of the molecule. In order to produce beams of species such as alkali metals or alkali metal halides it is necessary to heat the source to produce a sufficiently high vapour pressure. Effusive sources have three serious disadvantages for scattering studies. If one adheres to the conditions for effusive flow, the resulting beam is of very low intensity. A typical number density in the crossing region would be 10^{10} molecules cm^{-3}. The second disadvantage is that the molecules in the beam have a

Maxwell–Boltzmann distribution of velocities. Velocity selection can be achieved by using a mechanical device consisting of a set of slotted discs mounted on a common axis. For a particular velocity of rotation only those molecules having velocities within a narrow range are transmitted. However, use of such a device results in a substantial reduction in intensity and it is not feasible to use velocity selection for both beams. The third disadvantage is that one can only vary the velocity over a very limited range.

These disadvantages are overcome by the use of supersonic nozzle beams in which gas expands from a high pressure through a small nozzle into a vacuum. A skimmer removes excess gas which is pumped away. A schematic diagram of a nozzle source is given in figure 7.1.

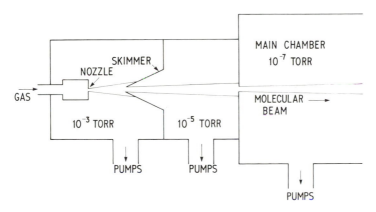

Figure 7.1. Schematic diagram of a molecular beam nozzle source.

Each of the two chambers through which the beam passes before entering the main vacuum chamber is separately pumped. This is known as differential pumping. This technique results in much more intense beams which have a relatively narrow velocity spread of the order of 10% of the most probable velocity. The velocity of the beam is also more readily variable than is the case with effusive sources. One can either vary the temperature of the source or use the technique of seeding. In seeded nozzle beams, a mixture of the reactant gas in an excess of an inert carrier gas is expanded through the nozzle. If the reactant is a heavy molecule and the carrier gas is light (e.g. helium or neon), the reactant molecules will have very nearly the velocity of the light carrier gas. Intermediate velocities can be obtained by using a mixture of inert gases as carrier gas and by this technique the translational energy can be varied from about 0.05 eV to about 5 eV. The expansion results in a cooling of the gas which is accompanied by rotational cooling. Another characteristic of supersonic nozzle sources is that in many cases the beam contains an appreciable proportion of van der Waals dimers. Thus one can investigate the dynamics and spectroscopy of .these weakly bound molecules, and this will be discussed in chapter 8. The use of supersonic seeded beams is not restricted to stable molecules. It is now possible to produce intense beams of atoms such as O and Cl and of radicals such as OH (Grice 1981b). A schematic diagram of a crossed molecular beam apparatus with nozzle sources is given in figure 7.2.

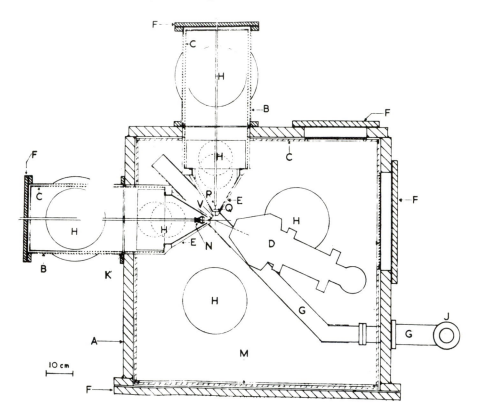

Figure 7.2. Schematic diagram of a molecular beam apparatus with a nozzle source. A, stainless steel scattering chamber; B, stainless steel source chambers; C, liquid nitrogen cooled cold shield; D, detector UHV chamber; E, source bulkheads; F, aluminium alloy flanges; G, liquid nitrogen cold trap; H, oil diffusion pumps; N, free radical source; P, nozzle source; Q, skimmer; V, beam chopper. (Reproduced from Carter *et al.*, *Farad. Disc. Chem. Soc.*, **55**, 357 (1973).)

Molecular beam methods can also be applied to ion–molecule reactions (Herman 1972, Koski 1975, Gentry 1979). Ions are most commonly produced by electron impact but chemical ionization in a high pressure source can also be used. This latter technique has the advantage that the ions are substantially vibrationally relaxed. The required ion is selected by using a mass spectrometer as a mass filter. The translational energy of the beam is readily varied by the use of acceleration or deceleration electric fields. The production of low-energy ion beams is difficult because of space charge effects but it is now possible to produce intense ion beams (of the order of 5×10^{-10} A to 1×10^{-9} A at the crossing region) having laboratory energies of a few tenths of an electron volt. A current of 10^{-9} A corresponds to about 6×10^{9} ions per second.

Thus it is possible to perform molecular beam scattering experiments in which the initial relative velocity is reasonably well defined, but it is less easy to select particular internal states of a molecular species. Beams containing a substantial proportion of vibrationally excited molecules can be produced by laser excitation. For example, the effect of vibrational excitation in the reaction

$$K + HCl \rightarrow KCl + H \tag{7.1}$$

has been demonstrated by exciting some of the HCl molecules to the $v = 1$ state (Odiorne *et al.* 1971). It is also possible to selectively populate particular rotational levels of the $v = 1$ state. However, only a small fraction of the ground state molecules are excited and this technique can be used only in cases where there is a strong dependence of the reaction cross-section on the vibrational state of the reactant molecule.

Particular rotational states for a given vibrational state can be selected by the use of inhomogeneous electric fields. The energy of a polar diatomic molecule in an electric field is given by a second-order Stark effect which depends on the square of the electric field and on the rotational quantum numbers J and $|M_J|$ (where M_J gives the component of J in the direction of the electric field). Molecules having particular values of J and $|M_J|$ can be focused by a quadrupole electric field. This property enables one to select, in a molecular beam experiment, particular rotational states. Spherical top molecules have a first-order Stark effect and by using a hexapole field one can focus molecules having a particular negative value for the combination of quantum numbers $KM_J/J(J+1)$ (where K gives the component of J along the principal axis of the molecule). This technique enables one to produce a beam of molecules in which there is a preferential orientation of the principal axis (Brooks 1976). These selection techniques do, of course, result in a substantial reduction in beam intensity and can only be applied to one beam in reactions which have a substantial reactive cross-section.

7.1.1.2. Molecular Beam Detectors and Velocity Analysis

Until 1969 molecular beam studies of reactive scattering were restricted to reactions involving alkali metals. The reason for this is that it is relatively easy to measure the intensity of scattered alkali metal atoms and molecules containing alkali metals, whereas it is much more difficult to detect non-alkali species. When an alkali metal atom or a molecule containing an alkali metal atom strikes a hot tungsten filament it is very efficiently ionized. The ions evaporate and can be counted. One can distinguish between alkali atoms and alkali halides by using two filaments. A tungsten filament measures the flux of both species but a filament of platinum containing about 8% of tungsten will detect only alkali atoms. Thus the flux of alkali halide molecules is given by the difference between the two measured fluxes.

Non-alkali species can be detected by electron bombardment of the scattered species and mass spectrometric detection of the resulting ions. Ionization effiency is low (about one molecule in every thousand is ionized) and it is necessary to have low background pressures in the detector (less than 10^{-13} torr).

In molecular beam studies of ion–molecule reactions, detection of ionic species is rather more straightforward because a mass spectrometer can be used directly to identify the scattered ion, and intensities can be measured by ion counting techniques.

In addition to measuring the intensity of the scattered species as a function of angle, it is also important to measure the velocity distribution. Indeed, measured velocity distributions are essential if one is to transform the data unequivocally from the laboratory coordinate system to the centre-of-mass system. For neutral species, the velocity distribution is measured by the time-of-flight method. A rotating chopper is mounted between the crossing region and the entrance to the detector. The intensity of the molecules reaching the detector is measured as a function of the time elapsed from the point at which the chopper starts to transmit product molecules. The disadvantage of the method is that if a disc with a single slit is used only about 5% of the scattered species is transmitted. This can be increased to 50% by the cross-correlation method which uses a disc with a series of pseudo-random slits and teeth and extracting the velocity distribution with the aid of a computer (Nowikow and Grice 1979).

The energy distribution of ionic products in ion–molecule reactions can be measured by the use of retardation potentials. The potential on a lens at the front of the detector is varied from zero to a value sufficient to prevent all product ions from reaching the detector. The intensity of ions transmitted is recorded as a function of the applied potential and numerical differentiation of this integral energy distribution yields the energy distribution of the ion in question.

7.1.1.3. Analysis of the Internal State of the Product Molecule: Laser-Induced Fluorescence

Since the intensity of the product molecules is very low, one cannot use conventional spectroscopic techniques to determine the internal state. In some reactions the product is frequently in an excited electronic state which will undergo transitions to the ground state. The spectrum of the chemiluminescence can be recorded and interpreted to give a detailed picture of the distribution among vibrational and rotational states (Menzinger 1980, Ottinger 1984). However, this technique is limited to reactants producing excited states that radiate.

A more general method for determining the internal state distribution of product molecules formed in molecular beam experiments is that of laser-induced fluorescence. This technique has been applied both to the crossed-beam experiment and to the beam-gas method. A beam of radiation from a tunable laser passes through the crossing region and the laser frequency is swept. As it passes through a frequency connecting a particular $v''J''$ state in the ground electronic state with a vibration–rotation state in an excited electronic state, excitation occurs. The excited molecule can then fluoresce and the undispersed fluorescence is measured as a function of the excitation frequency. The spectrum thus indicates which product states $v''J''$ are formed in the reaction. In order to obtain relative populations for these various states, it is necessary to know the Franck–Condon factors and rotational line strengths for the transitions involved. The method has the advantage of being applicable to ground-state products but the molecule must have a strong electronic absorption band in the spectral ranges covered by available lasers. One needs a fairly detailed knowledge of the spectroscopy of the molecule in question and it must have an appreciable fluorescence quantum yield. The majority of studies have been made with alkaline earth halides in reactions such as

$$Ba + HF \rightarrow BaF + H \tag{7.2}$$

(Cruse *et al.* 1973, Zare and Dagdigian 1974, Zare 1979). Alkaline earth halides have several strong electronic transitions in the visible region of the spectrum. A diagram of a crossed-beam apparatus with laser-induced fluorescence detection is shown in figure 7.3.

7.1.2. Transformation of Molecular Beam Data from Laboratory to Centre-of-Mass Coordinates

In a crossed-beam experiment, the two reactant beams usually intersect at right angles. The velocities of the individual reactants and of one of the products are measured and the intensity of the scattered product is obtained as a function of the scattering angle with respect to the laboratory coordinates. In the absence of external fields, the velocity of the centre-of-mass is invariant and is unimportant from the point of view of the reaction dynamics. In order to interpret the data it is necessary to transform the velocities and angles to a centre-of-mass coordinate system. This is done using a Newton diagram as shown in figure 7.4.

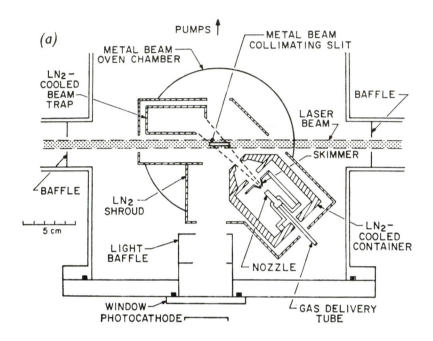

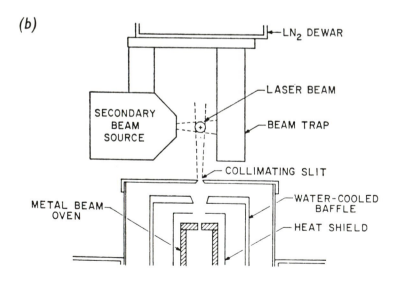

Figure 7.3. A crossed molecular beam apparatus with laser-induced fluorescence detection. (Reproduced from Cruse *et al.* 1973.)

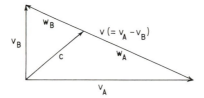

Figure 7.4. Newton diagram showing transformation of laboratory velocities v_A and v_B to centre-of-mass velocities w_A and w_B.

The vectors v_A and v_B represent the laboratory velocities of reactants A and B. The initial relative velocity vector v is given by $v_A - v_B$ as shown in the figure and has magnitude $(|v_A|^2 + |v_B|^2)^{1/2}$. If the velocity of the centre-of-mass is represented by the vector c, conservation of momentum requires that

$$(m_A + m_B)c = m_A v_A + m_B v_B, \qquad (7.3)$$

where m_A and m_B are the masses of species A and B. The velocity of the centre of mass is constant and does not change in the collision. The velocities of A and B relative to the centre of mass, w_A and w_B respectively, are then given by

$$w_A = v_A - c = v_A - \frac{(m_A v_A + m_B v_B)}{m_A + m_B} = \frac{m_B v}{m_A + m_B}, \qquad (7.4)$$

$$w_B = v_B - c = - \frac{m_A v}{m_A + m_B}. \qquad (7.5)$$

Clearly $v = w_A - w_B$. After the scattering event, the product molecules C and D will have laboratory velocities v_C and v_D and centre-of-mass velocities w_C and w_D. Expressions similar to those of equations 7.4 and 7.5 relate w_C with v_C and w_D with v_D, namely,

$$w_C = v_C - c = \frac{m_D v'}{m_C + m_D}, \qquad (7.6)$$

$$w_D = v_D - c = - \frac{m_C v'}{m_C + m_D}, \qquad (7.7)$$

where v' is the final relative velocity vector

$$v' = v_C - v_D = w_C - w_D. \qquad (7.8)$$

In elastic scattering, the magnitude of the relative velocity is unchanged (i.e. $|v'| = |v|$) but its direction changes. In order to relate the final relative velocity in an inelastic or reactive collision to the initial relative velocity, we have to equate the initial and final energies

$$E'_T + E'_I = E_T + E_I - \Delta E_0, \tag{7.9}$$

where E'_T and E_T are the final and initial relative translational energies, E'_I and E_I are the final and initial internal energies and ΔE_0 is the difference between the zero-point energies of the reactants and products. The final relative velocity v' is related to E'_T by

$$|v'| = \left(\frac{2E'_T}{\mu'}\right)^{1/2}, \tag{7.10}$$

where μ' is the reduced mass of the product, $m_C m_D / (m_C + m_D)$. The laboratory scattering angle Θ for species C is the angle between the vector v_C and v_A, whereas the centre-of-mass scattering angle θ is the angle between w_C and the initial relative velocity vector v (figure 7.5).

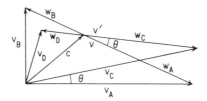

Figure 7.5. Newton diagram showing relationship of reactant and product laboratory velocities to centre-of-mass velocities.

Since the final internal energy can take a range of values from zero to the dissociation energy D_0 of the molecule, the final relative kinetic energy can vary from a maximum value of $E_T + E_I - \Delta E_0$ to a lower limit of $E_T + E_I - \Delta E_0 - D_0$. The magnitude of the final relative velocity will thus lie between a maximum value corresponding to no internal excitation of the product, and a minimum value corresponding to the dissociation limit. Similarly the magnitude of the final velocity of a molecular product, say C, will lie between two limits. For a given internal excitation E'_I, the velocity vector of C will pivot about an origin at the centre-of-mass and the tip of the vector will sweep out a sphere. We can restrict our discussion to two dimensions because of the cylindrical symmetry with respect to rotation about the initial relative velocity vector v as discussed in section 6.1 and represent the magnitude of the velocity of C by a circle as in figure 7.6. If we assume that the vector v' (and hence w_C) can have any orientation with respect to v, then for a given magnitude of w_C, the corresponding laboratory velocity vector v_C will be confined to a cone about the velocity of the centre of mass, c (figure 7.7). This leads to an ambiguity in laboratory measurements if velocity analysis is not employed, because product molecules with two different

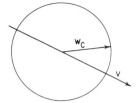

Figure 7.6. Distribution of product velocity w_C with respect to centre-of-mass.

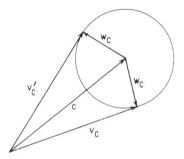

Figure 7.7. Range of laboratory velocities v_C resulting from a given magnitude of w_C.

magnitudes and directions of w_C (corresponding to two different degrees of internal excitation) can contribute to scattering at a particular laboratory angle (figure 7.8).

The results of molecular beam scattering experiments are usually presented in the form of intensity contour diagrams. The centre-of-mass of the system is taken as the origin and the direction of the initial relative velocity vector v corresponds to zero deflection. The angle of scattering is plotted with respect to this direction and the velocity of the scattered particle is represented by the distance of the point on the diagram from the origin. The intensity distribution is then represented by contours joining points of equal intensity. An example is shown in figure 7.9 for the distribution of KI from the reaction

$$K + I_2 \rightarrow KI + I \tag{7.11}$$

(Gillen *et al.* 1971). The maximum intensity is seen to be in the direction of the initial relative velocity at a velocity of $6 \times 10^4 \, \text{cm s}^{-1}$. Smaller amounts of KI are scattered at other angles and in the 0° direction with velocities smaller than or larger than $6 \times 10^4 \, \text{cm s}^{-1}$. The interpretation of such diagrams will be discussed in subsequent sections.

In order to plot an intensity contour diagram it is necessary first of all to transform the laboratory intensity data, in the form of the relative differential

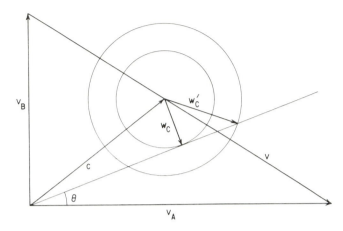

Figure 7.8. Contribution to scattering at laboratory angle Θ of products with centre-of-mass velocities w_C and w'_C.

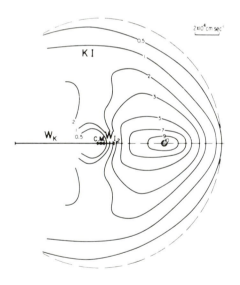

Figure 7.9. Product intensity contour diagram for KI produced in reaction of K with I_2. (Reproduced from Levine and Bernstein 1974, adapted from Gillen *et al.* 1971.)

cross-section as a function of laboratory angle Θ and laboratory velocity $|v_C|$, into the centre-of-mass differential cross-section in terms of the scattering angle θ with respect to v, the initial relative velocity vector, and velocity $|w_C|$ of product C with respect to the centre-of-mass. In the case of elastic scattering, the centre-of-mass velocities are unchanged and the transformation is given by

$$d\sigma_{CM}(\theta) = \frac{|\boldsymbol{w}_C|^2}{|\boldsymbol{v}_C|^2}|\cos\delta|\,d\sigma_{LAB}(\Theta), \tag{7.12}$$

where $d\sigma_{CM}(\theta)$ and $d\sigma_{LAB}(\Theta)$ are the differential cross-sections in the centre-of-mass and laboratory coordinate systems respectively and δ is the angle between the vectors $\boldsymbol{w}_C$ and $\boldsymbol{v}_C$.

In the case of reactive scattering, the velocities of the products can cover a wide range of values and the transformation between the laboratory differential cross-section (or measured flux) and the centre-of-mass differential cross-section (or flux) is given by

$$d\sigma_{CM}(\theta, w_C) = \frac{w_C^2}{v_C^2}d\sigma_{LAB}(\Theta, v_C), \tag{7.13}$$

where w_C and v_C are the magnitudes of $\boldsymbol{w}_C$ and $\boldsymbol{v}_C$, respectively. For details of the transformation, the reader is referred to Bernstein (1982), Warnock and Bernstein (1968), and Wolfgang and Cross (1969).

However, there is a spread of velocities in each reactant beam and there is thus no unique centre-of-mass velocity. It is necessary to perform a complicated deconvolution calculation (Siska 1973) to obtain the centre-of-mass differential cross-sections in the polar coordinate system (θ, w_C). An alternative presentation uses a Cartesian coordinate system and yields probabilities of the product having velocity components in the range w_x to $w_x + dw_x$, w_y to $w_y + dw_y$ and w_z to $w_z + dw_z$. The probability functions are related to the differential cross-sections by the relationship

$$P_{LAB}(v_x, v_y, v_z) = P'_{CM}(w_x, w_y, w_z) = \frac{d\sigma_{LAB}(\Theta, v)}{v^2} = \frac{d\sigma_{CM}(\theta, w)}{w^2}. \tag{7.14}$$

(Wolfgang and Cross 1969, Herman and Wolfgang 1972).

7.2. Determination of Interatomic Potentials by Elastic Scattering

In chapter 2 we were concerned with potential curves for diatomic molecules in which two atoms are bound together by valence forces. Atoms which do not interact to form bound molecules do exert intermolecular forces upon each other and one can define an interatomic potential in much the same way as for a chemically bound species. In the case of two atoms, the potential will have the same general form as a diatomic potential (as shown in figure 2.1) but the depth of the well will be much smaller and the potential minimum will occur at a much larger interatomic separation. For Ar...Ar, the well depth ϵ is approximately $0.012\,\text{eV}$ and the internuclear separation at the minimum R_m is $3.76\,\text{Å}$. In this section we shall be concerned with how information about the interatomic potential can be derived from elastic scattering studies.

Potential Energy Surfaces

We will first consider elastic scattering from a fairly qualitative point of view. In figure 6.2 we considered trajectories with impact parameters in the range b to $b+db$ and argued that all such trajectories would be scattered into an annular cone defined by the angles θ and $\theta+d\theta$. The differential cross-section $d\sigma'(\theta, v)$ was related to the impact parameter b by the equation

$$d\sigma'(\theta, v)\sin\theta\, d\theta = b\, db. \tag{7.15}$$

In elastic scattering v is unchanged and we can rewrite the equation as

$$d\sigma(\theta) = \frac{b}{\sin\theta \dfrac{d\theta}{db}}. \tag{7.16}$$

The derivative $d\theta/db$ is related to the deflection function which expresses the angle through which the trajectory is scattered in terms of the impact parameter b. This is illustrated in figure 7.10, where we regard one particle as being stationary at the origin O and the other particle, with an effective mass equal to the reduced mass, μ, approaching with a relative velocity v.

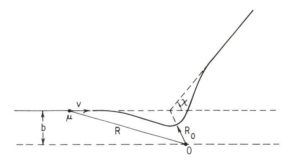

Figure 7.10. Elastic scattering of a particle with effective mass μ and initial relative velocity v by a stationary particle at the origin O.

The deflection χ of the trajectory is given by the classical deflection function

$$\chi = \pi - 2b \int_{R_0}^{\infty} \frac{dR}{R^2}\left(1 - \frac{V(R)}{E} - \frac{b^2}{R^2}\right)^{1/2}, \tag{7.17}$$

where R is the distance between the particles, R_0 is the distance of closest approach, E is the relative kinetic energy and $V(R)$ is the interatomic potential. Since χ can be positive or negative, the observed scattering angle $\theta = |\chi|$ and

$$d\sigma(\theta) = \frac{b}{\sin\theta \left| \dfrac{d\chi}{db} \right|} .$$ (7.18)

This equation connects the measured differential cross-section with the interatomic potential. Let us now consider qualitatively how χ varies as b increases from zero. If b is zero, the collision will be head on and the deflection angle will be $180°$. As b increases, the interaction becomes less repulsive and χ will steadily decrease from $180°$. There will be a certain value of b for which the attractive component of the potential will cancel out the repulsive contribution resulting in zero deflection. Such trajectories are known as glory trajectories and we will discuss the significance of these later. As b increases further, the attractive contribution becomes dominant resulting in negative values of χ. However, the decrease in χ does not continue indefinitely because if the two particles are drawn too close to each other, the repulsive force starts to operate and χ increases. Thus there is a minimum value of χ, χ_R, beyond which deflection cannot occur. There is a bunching up of intensity near this angle and because of the analogy with the optical rainbow formed by the refraction of light rays by drops of water, this minimum angle is known as the rainbow angle (Mason *et al.* 1971). As b increases further, χ increases steadily towards zero for large b. Figure 7.11 illustrates qualitatively trajectories for different impact parameters and the deflection function is shown in figure 7.12. At the rainbow angle $d\chi/db$ is zero and there will be a singularity in the differential cross-section as given by equation 7.18. Similarly there will be a singularity at $\theta = 0$ and the differential cross-section has the form shown in figure 7.13.

Although classical mechanics is a useful starting point for the discussion of elastic scattering, quantum effects are important. We see from figure 7.12 that for some ranges of θ there will be three values of b giving trajectories with the same value of θ. Thus quantum mechanical interference can occur resulting in widely spaced oscillations in the differential cross-section. These oscillations are known as supernumerary rainbows. The singularity at the rainbow angle is also removed and the differential cross-section has the form shown in figure 7.14. Differential cross-sections for the elastic scattering of Na with Hg are shown in figure 7.15 (Buck and Pauly 1971) where several supernumerary rainbows are clearly visible. The spacing and shape of the supernumerary rainbows depends on the shape of the intermolecular potential. Superimposed on these widely spaced oscillations are rapid secondary oscillations arising from a quantum mechanical diffraction phenomenon. This is illustrated in the results of quantum scattering calculations shown in figure 7.16.

In cases where five or more extrema (maxima or minima) can be resolved in the supernumerary rainbow structure of the experimental differential cross-section it is possible to determine the interatomic potential by

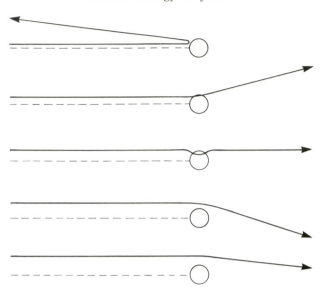

Figure 7.11. Trajectories for different impact parameters.

inversion of the deflection function $\chi(b)$. It is difficult to invert a function such as that shown in figure 2.1 which is not monotonic. The deflection function is written as a sum of functions $g_i(b)$

$$\chi(b) = \sum_i g_i(b). \tag{7.19}$$

The functions $g_i(b)$ contain variable parameters which are chosen to reproduce the observed differential cross-section. The data for this consist of the positions of the maxima and minima in the supernumerary rainbows, the amplitude of the rainbow oscillations, the separation of the rapid quantum oscillations, the monotonic large angle scattering, the positions of the glory oscillations in the integral cross-section (see below) and the coefficient C_6 of the R^{-6} term which is usually used to describe the attractive part of the potential. Analytic classical deflection functions are assumed for three regions, namely $\theta \approx \pi, b = 0; \theta \approx \theta_R, b = b_R$ and $\theta_R < \theta < 0, b > b_R$ and some parameters are chosen to ensure continuity of the function in equation 7.19. It is then possible to use a semi-classical method to calculate the potential function for a substantial range of interatomic separations from the deflection function $\chi(b)$. Details of the inversion procedure are given by Buck (1974, 1975) and by Maitland *et al.* (1981). The potential function is not constrained to fit any particular analytical function and is therefore a better representation than potentials derived by adjusting parameters in a particular model function to fit experimental data. However, the inversion

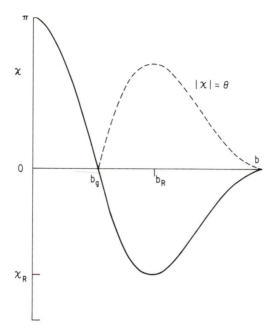

Figure 7.12. Deflection function $\chi(b)$ for elastic scattering.

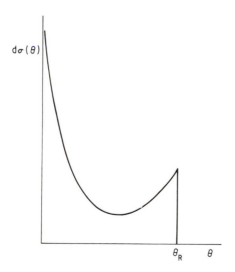

Figure 7.13. Classical differential cross-section for elastic scattering.

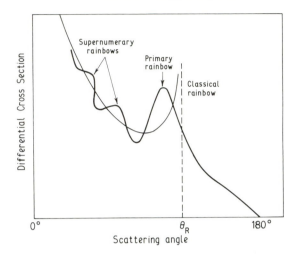

Figure 7.14. Differential cross-section for elastic scattering showing primary rainbow and supernumerary rainbows.

technique is only applicable to potentials with deep wells for systems with heavy atoms. The spacing between the secondary oscillations is given by the approximate formula

$$\Delta\theta \approx h/2\mu v R_{\mathrm{m}}, \tag{7.20}$$

where μ is the reduced mass and v is the relative velocity. Thus if secondary oscillations can be resolved, R_{m} can be determined.

In cases where the data are not sufficiently accurate for the calculation of the potential by inversion of the deflection function, one can still make certain deductions about the potential function from the differential cross-section. However, in such cases a particular functional form is assumed and although it may give a very good fit to some regions of the potential, it may give a poor description in other regions. If the impact parameter is relatively large, the scattering is dominated by the long range attractive part of the potential which is usually taken to be of the form $-C_s R^{-s}$ where s is an integer. The differential cross-section is given by

$$\mathrm{d}\sigma(\theta) = f(s)\left(\frac{C_s}{E}\right)^{2/s} \theta^{-(2+s)/s}, \tag{7.21}$$

where E is the relative kinetic energy and $f(s)$ is a function of s. The parameters s and C_s can be obtained from the slope and intercept of a plot of $\log \mathrm{d}\sigma(\theta)$ against $\log \theta$. Also the position of the rainbow angle, θ_{R}, is proportional to ϵ/E (where ϵ is the well depth) and can be used to estimate the well depth. At higher energies the scattering is dominated by the repulsive part

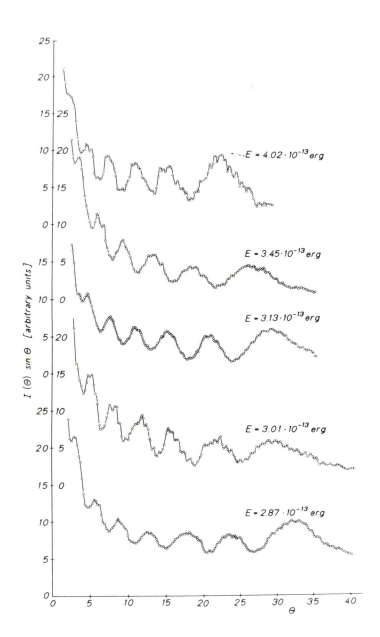

Figure 7.15. Experimental differential cross-sections for the elastic scattering of Na with Hg. (Reproduced from Buck and Pauly 1971.)

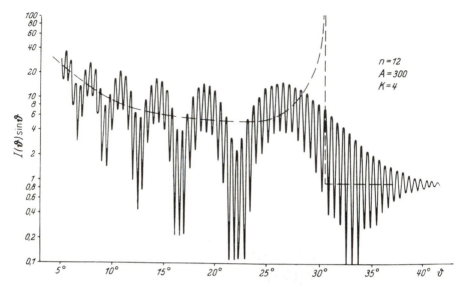

Figure 7.16. Differential cross-section calculated by quantum scattering theory. (Reproduced from Hundhausen and Pauly, *Z. für Physik*, **187**, 305 (1965).)

of the potential and the attractive part plays virtually no part. Under these conditions one can regard the potential as being monotonic and the deflection function can be calculated readily from the differential cross-section. The potential function is then obtained by inverting the deflection function.

Useful information can also be obtained from integral cross-section measurements obtained from beam–gas experiments in which one measures the attenuation of the intensity of a beam of species A on passing through a collision chamber containing species B. The attenuation of the beam is related to the total scattering cross-section σ by the equation

$$I(x) = I(0)\exp(-n_B \sigma x) \qquad (7.22)$$

(from section 6.1), where $I(0)$ is the incident flux and $I(x)$ is the flux emerging from a scattering chamber of length x containing gas B with number density n_B. For low relative velocities, the scattering is dominated by the attractive part of the potential. As the relative energy is increased one would expect, on qualitative grounds, that the well in the potential would have a decreasing effect on the scattering and that the integral scattering cross-section should decrease. If the attractive part of the potential can be represented by a term of the type $C_n r^{-n}$ then the integral cross-section $\sigma(E)$ is given by the equation

$$\sigma(E) \approx f(n)\left(\frac{\pi\mu C_n}{nE}\right)^{2/(n-1)}, \tag{7.23}$$

where μ is the reduced mass, E is the relative kinetic energy and $f(n)$ is a function of n. The power n can be determined from the velocity-dependence of $\sigma(E)$ and if the absolute cross-section is measured then C_n can be determined. However, absolute cross-section measurements are very difficult to make. At higher energies the scattering is dominated by the repulsive part of the potential and if this can be represented by a term $C_s r^{-s}$, C_s can again be obtained from the velocity-dependence of the integral cross-section by use of equation 7.23.

We mentioned above that there is a particular value of the impact parameter for which the attractive and repulsive contributions of the potential exactly balance resulting in zero deflection of the scattered particle. Zero deflection also occurs for values of the impact parameter which are so large that the potential has no influence on the trajectory of the particle. Quantum mechanical interference can occur between these two types of trajectory. This is known as the glory effect, by analogy with the optical glory effect, and results in oscillations in the variation of the integral cross-section with relative velocity. For the case of the Lennard-Jones potential function (see section 8.1.3)

$$V(R) = \frac{C_{12}}{R^{12}} - \frac{C_6}{R^6}, \tag{7.24}$$

the variation of integral cross-section with velocity is as shown in figure 7.17. The spacing of the oscillations depends on the product, ϵR_m, of the well depth ϵ and the internuclear distance, R_m at the potential minimum.

7.3. Molecular Beam Studies of Chemical Reactions

7.3.1. *Direct Reactions and Reactions Proceeding via Collision Complexes*

In the case of reactive scattering it is not possible to proceed directly from the experimental differential cross-section to the potential surface. One can only make deductions about the nature of the potential surface from the observed scattering. Thus the results of trajectory calculations, as outlined in chapter 6, are of great value in the interpretation of the dynamics. Many reactions occur within a very short time ($\leqslant 10^{-13}$ s) and are termed direct whereas in other cases there is evidence of a complex with a lifetime of more than a few rotational periods. If the reaction is direct then the product intensity contour diagram will be asymmetric with the contours of maximum

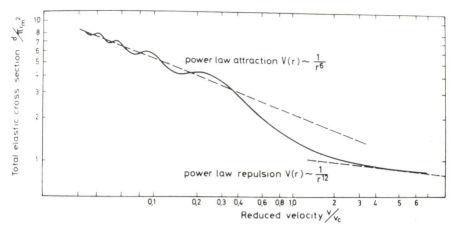

Figure 7.17. Glory oscillations in the total elastic cross-section as a function of relative velocity. (Reproduced from Toennies 1974.)

intensity in the direction of the relative velocity vector of the particular product molecule being considered.

However, if a collision complex is formed which has a lifetime of a few rotational periods before it splits up into products, the direction of the final velocity vector will be random. Let us consider a simplified system in which both collision partners are atoms. The diatomic complex will have an angular momentum l which will be perpendicular to the initial relative velocity vector $\boldsymbol{v}$. The magnitude of l will be given by

$$|l| = \mu b |\boldsymbol{v}|, \qquad (7.25)$$

where μ is the reduced mass and b is the impact parameter (figure 7.10). In the absence of external fields, the principle of conservation of momentum requires the rotational motion to be confined to a plane defined by $\boldsymbol{v}$ and b. The final velocity vector $\boldsymbol{v}'$ can rotate in this plane, which is perpendicular to l, and thus for random decomposition of the complex the final relative velocity vector can have any orientation in this plane (figure 7.18). At first sight one might expect the distribution of $\boldsymbol{v}'$, and of the product, to be isotropic but this ignores the cylindrical symmetry of the scattering with respect to the axis defined by $\boldsymbol{v}$. One really has to consider a three-dimensional situation obtained by the rotation of figure 7.18 about the vector $\boldsymbol{v}$. The final velocity vector $\boldsymbol{v}'$ (and the centre-of-mass velocity vectors of the products) will be distributed over the surface of a sphere. However, this distribution will not be isotropic. If a detector at an angle θ to $\boldsymbol{v}$ measures a flux $I(\theta)$, the number of product molecules scattered through an angle θ is proportional to $I(\theta)d\omega$, where $d\omega$ is the element of solid angle subtended by an annulus defined by θ and $\theta + d\theta$ (as discussed in section 6.1). Thus the number of particles

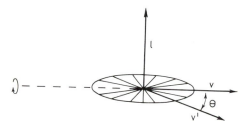

Figure 7.18. Relationship between the final relative velocity v' and the initial relative velocity v for a diatomic complex.

scattered through an angle θ is proportional to $I(\theta)\sin\theta \, d\theta \, d\phi$. For random decomposition of the complex we require $I(\theta)\sin\theta \, d\theta \, d\phi$ to be constant and thus the measured flux $I(\theta)$ will be proportional to $(\sin\theta)^{-1}$ and thus will be peaked at $\theta = 0°$ and $\theta = 180°$. On the basis of this simplified model, the product intensity contour diagram for a reaction proceeding via a collision complex will be symmetrical with respect to the centre-of-mass scattering angle $\theta = 90°$ and will have maximum intensity at $\theta = 0°$ and $\theta = 180°$.

The situation is rather more complicated for the case of an atom A reacting with a diatomic molecule BC because we now have to consider the rotational angular momentum j of BC as well as the orbital angular momentum l of A relative to BC. The total angular momentum J is conserved and thus

$$J = l + j = l' + j', \tag{7.26}$$

where l' and j' are the orbital and rotational angular momenta of the products. The vector J is not perpendicular to v (figure 7.19) and hence the final velocity vector v' is no longer confined to the plane perpendicular to l. However, the magnitude of l is usually very much larger than that of j and thus $J \approx l$. The scattering still has cylindrical symmetry with respect to v. In these circumstances the intensity distribution will still be symmetric with respect to $\theta = 90°$ but will not necessarily have peaks in the forward and backward directions. Sideways peaking is expected to occur when a complex which is an oblate symmetric top dissociates along its symmetry axis. For a fuller discussion the reader is referred to Miller *et al.* (1967).

7.3.2. Direct Reactions

7.3.2.1. The Stripping Mechanism

The most distinctive feature in intensity contour diagrams for direct reactions is whether the intensity maximum is in the forward direction (at $0°$

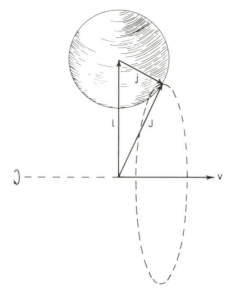

Figure 7.19. Angular momenta for a complex formed from an atom and a diatomic molecule.

with respect to the initial relative velocity vector $\boldsymbol{v}$) or in the backward direction (at 180° with respect to $\boldsymbol{v}$). In the intensity contour diagram for the reaction

$$K + I_2 \rightarrow KI + I \qquad (7.27)$$

(Gillen *et al.* 1971) shown in figure 7.20, the product KI is scattered forwards. Such reactions, for which the scattering is predominantly in the forward direction, are said to occur by a stripping mechanism. The appearance of the product KI at 0° indicates that, in the centre-of-mass coordinate system, the potassium is not deflected during the reactive collision. This is interpreted in terms of a grazing collision occurring with a relatively large impact parameter as indicated in figure 7.20. The remaining iodine atom, which becomes the product I, is virtually unaffected by the collision. If we make the extreme assumption that in a reaction

$$A + BC \rightarrow AB + C \qquad (7.28)$$

the velocity of the product atom C is unchanged, we can make a simple calculation for the velocity of the molecular product AB. This very simple model, in which there is no transfer of momentum to atom C, is known as the spectator stripping model. Applying the principle of conservation of momentum gives

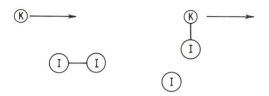

Figure 7.20. Grazing collision between K and I_2 resulting in spectator stripping.

$$m_A w_A + m_{BC} w_{BC} = m_{AB} w_{AB} + m_C w_C. \qquad (7.29)$$

Since w_C is assumed to be the same as w_{BC}, we can rearrange the equation to yield an expression for w_{AB}:

$$w_{AB} = \frac{m_A w_A + m_{BC} w_{BC} - m_C w_C}{m_{AB}} = \frac{m_A w_A + m_B w_{BC}}{m_{AB}}. \qquad (7.30)$$

The centre-of-mass velocities w_A and w_{BC} can be expressed in terms of the initial relative velocity v by using equations 7.4 and 7.5 and substitution into equation 7.30 yields the spectator stripping velocity

$$w_{AB} = \frac{(m_A + m_C)v}{(m_A + m_B)(m_A + m_B + m_C)}. \qquad (7.31)$$

If the product has a most probable velocity which is larger than that predicted by this simple model, there must be a repulsive force between the products. Velocities lower that the spectator stripping velocity are indicative of more internal excitation than the model would predict.

Reactions of alkali metals with diatomic halogens and interhalogen compounds show strong forward peaking and proceed by a stripping mechanism. The relative translational energies of the products are low, indicating that most of the exothermicity of the reaction goes into internal excitation. The reactive cross-sections for reactions of alkali metal atoms with diatomic halogens are very much larger than the collision cross-sections. This is interpreted in terms of the 'harpoon' or 'electron jump' mechanism. We illustrate this by considering qualitatively potential surfaces for the reaction of $K + Br_2$. The entrance channel for the collinear $K + Br_2$ surface will be virtually horizontal at first and described by an intermolecular potential of the form $-C_6/R_x^6$, where R_x is the separation between the K and Br_2. However, this surface will be cut by the ionic surface $K^+ + Br_2^-$ (figure 7.21) which can be represented by a Coulombic potential $-e^2/4\pi\epsilon_0 R_x + \Delta E$, where ΔE is the difference in energy between the asymptotes $K + Br_2$ and $K^+ + Br_2^-$. This is equal to the difference between the ionization energy of K and the electron affinity of Br_2. The two surfaces would be expected to cross when

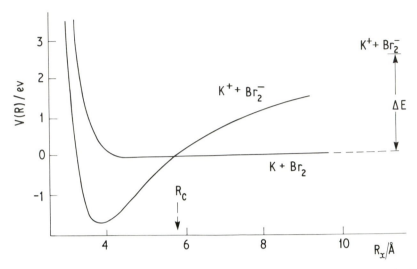

Figure 7.21. Schematic diagram of intersecting potential curves for $K + Br_2$ and $K^+ + Br_2^-$.

$$-\frac{e^2}{4\pi\epsilon_0 R_x} + \Delta E = -\frac{C_6}{R_x^6}. \qquad (7.32)$$

Since the intermolecular term on the right-hand side is negligible compared with the term on the left-hand side, the two surfaces will intersect at a value of R_x approximately equal to $e^2/4\pi\epsilon_0 \Delta E$. In the harpoon mechanism the transfer of an electron from K to Br_2 occurs in this region. There is a strong Coulombic attraction between K^+ and Br_2^- leading to a reactive cross-section which is much larger than the collision cross-section. The transferred electron is the harpoon which is pulled in by the Coulombic attraction. If we assume that b_{max}, the maximum value of the impact parameter for which reaction can occur, is equal to R_c, the value of R_x at which the surfaces intersect, then the reactive cross-section will be approximately equal to πR_c^2. The electron transferred to Br_2 enters an antibonding orbital and hence facilitates the breaking of the Br–Br bond. The strong attractive force between K^+ and Br_2^- offsets the repulsive Br–Br interaction and the product is scattered forwards.

Stripping dynamics have also been observed in reactions of halogen atoms with diatomic halogen molecules such as Cl_2 and Br_2, but here the electron jump mechanism is not applicable and the reactive cross-sections are much smaller than the collision cross-sections. Some oxygen atom reactions also proceed by stripping mechanisms (Grice 1981b). For example, in the reaction

$$O + CS_2 \rightarrow OS + CS \qquad (7.33)$$

the OS product is strongly forward scattered (figure 7.22).

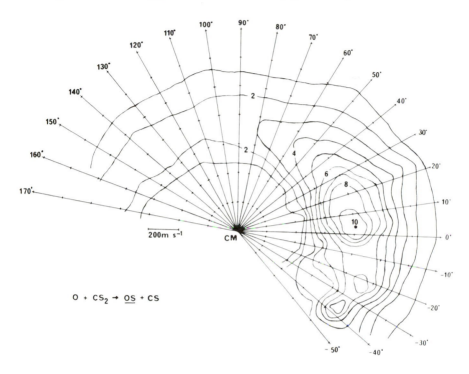

Figure 7.22. Product intensity contour diagram for OS produced in the reaction $O + CS_2$ showing forward scattering of OS. (Reproduced from Gorry *et al.*, *Mol. Phys.*, **37**, 329 (1979).)

However, at lower translational energies there is some evidence for the participation of a complex in the reaction of O with Br_2. The stripping mechanism is also found in many ion–molecule reactions. Figure 7.23 illustrates data, plotted in terms of Cartesian flux, for the reaction

$$N^+ + H_2 \rightarrow NH^+ + H \qquad (7.34)$$

(Hansen *et al.* 1980) at a relative initial energy of 3.60 eV. The maximum intensity is clearly in the forward direction but the most probable velocity is considerably less than the spectator stripping velocity.

7.3.2.2. Rebound Mechanism

The other extreme case for direct reactions is that for which the product intensity contour diagram has a maximum in the backward direction

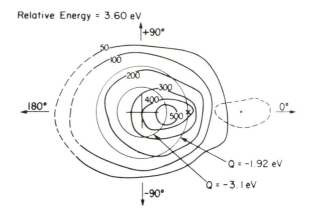

Figure 7.23. Product intensity contour diagram for NH^+ produced in the reaction of N^+ with H_2 at a relative energy of 3.60 eV. The spectator stripping velocity is marked with a cross. (Reproduced from Hansen *et al.* 1980.)

($\theta = 180°$ with respect to the relative velocity vector v). Thus the most probable velocity of the product is in the opposite direction to that of v. The distribution of KI from the reaction

$$K + CH_3I \rightarrow KI + CH_3 \qquad (7.35)$$

indicates that the KI is scattered backwards (Rulis and Bernstein 1972) (figure 7.24). This behaviour can be readily interpreted in terms of collisions in which the potassium approaches the CH_3I along the IC axis with a low impact parameter. If b_{max}, the maximum impact parameter for which reaction occurs, is of the order of atomic radii then scattering will be in the backward direction. For this reaction b_{max} is of the order of 3 Å, whereas for the reaction of K with Br_2 it is about 7 Å. The cross-sections for rebound reactions such as reaction 7.35 are much smaller than than those for stripping reactions between alkali atoms and halogen molecules. Another difference is that in rebound reactions most of the exothermicity is converted into translational energy showing that there is a strong repulsive force as the bond in the molecule being attacked breaks. The dynamics of reaction 7.35 can also be discussed in terms of an electron jump mechanism in which an electron is transferred from K to an antibonding orbital in CH_3I to form CH_3I^-, which dissociates rapidly. The CH_3I^- potential surface is strongly repulsive and we can therefore understand why a large fraction of the available energy appears as translational energy.

The product intensity contour diagram for the reaction

$$H + Cl_2 \rightarrow HCl + Cl \qquad (7.36)$$

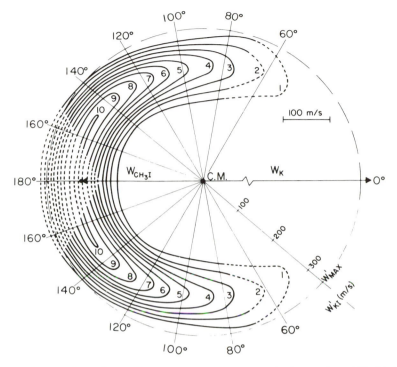

Figure 7.24. Product intensity contour diagram for KI produced in the reaction of K with CH_3I showing backward scattering of KI. (Reproduced from Rulis and Bernstein 1972.)

also shows strong backward scattering (figure 7.25) which is indicative of a collinear approach of H to Cl_2 with low impact parameter. The repulsive nature of the $H–Cl_2$ interaction can be understood in terms of a qualitative molecular orbital description. As H approaches Cl_2, the 1s orbital on H interacts with the antibonding σ orbital of Cl_2 to give an orbital which is H–Cl bonding but Cl–Cl antibonding. Thus breaking of the Cl–Cl bond with repulsive energy release to give HCl is favoured. Product intensity contour diagrams for the reactions of D with Br_2, I_2, ICl and IBr show pronounced sideways peaking for values of $\theta > 90°$ (McDonald *et al.* 1972) (figure 7.26). The reactive cross-sections are quite small indicating that the D atom has to approach quite close to the halogen molecule. There is substantial product translational energy release which occurs when the halogen bond breaks. The sideways peaking can be interpreted in terms of an increasingly bent geometry for DXY as the electronegativity of X decreases.

The results of a crossed molecular beam study of the reaction

$$F + D_2 \rightarrow DF + D \tag{7.37}$$

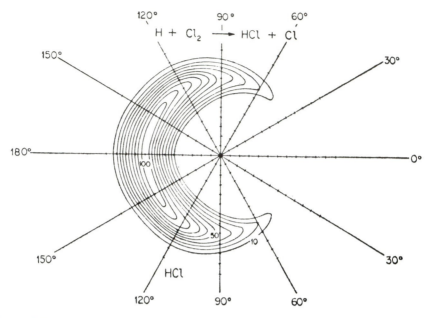

Figure 7.25. Product intensity contour diagram for HCl produced in the reaction $H + Cl_2$ showing backward scattering of HCl. (Reproduced from Herschbach 1973.)

are particularly interesting (Schafer *et al.* 1970). The vibrational levels of DF are widely separated and only relatively few rotational levels are populated. Hence the most probable velocities (with respect to the centre-of-mass) for the formation of DF in different vibrational states v' are widely spaced. The highest accessible level is for $v' = 4$. The product intensity contour diagram (figure 7.27), which indicates backward scattering of DF, also shows clear intensity peaks corresponding to the excitation of different v' levels of DF.

7.3.3. Reactions Proceeding via a Complex

In the reactions of alkali metal atoms with halogen molecules or alkyl halides, there is a transfer of an electron from the alkali atom to the reagent molecule. The transferred electron goes into an orbital which is antibonding and the breaking of the bond in the molecule is facilitated. When alkali metals react with molecules such as $SnCl_4$ or SF_6, the transferred electron goes into an orbital which is non-bonding or only weakly antibonding. A collision complex is thus formed and if its lifetime is more than a few rotational periods, the product intensity contour diagram will be symmetric with respect to the centre-of-mass scattering angle $\theta = 90°$. An example is shown

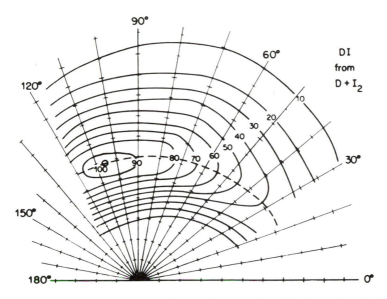

Figure 7.26. Product intensity contour diagram for DI produced in the reaction $D + I_2$ showing sideways peaking of DI. (Reproduced from McDonald *et al.* 1972.)

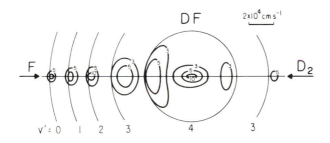

Figure 7.27. Product intensity contour diagram for DF produced in the reaction $F + D_2$ showing separate peaks for different vibrational states of DF. (Reproduced from Levine and Bernstein 1974, adapted from Schafer *et al.* 1970.)

in figure 7.28 for CsF formed in the reaction of Cs with SF_6 (Riley and Herschbach 1973). It is reasonable that the $CsSF_6$ complex should have a long lifetime because it has a large number of degrees of freedom and the anion SF_6^- has high bond energies.

The intensity contour diagram for C_2H_3F produced in the reaction of fluorine atoms with C_2H_4 (Parson *et al.* 1973) (figure 7.29) is clearly symmetric with respect to $\theta = 90°$ but is sideways peaked with the maximum intensity at 90°. However, the collision complex is a prolate top and not an oblate top for which sideways peaking would be expected. In this case the

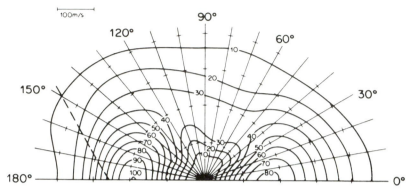

Figure 7.28. Product intensity contour diagram for CsF produced in the reaction $Cs + SF_6$ showing substantial forward–backward symmetry. (Reproduced from Riley and Herschbach 1973.)

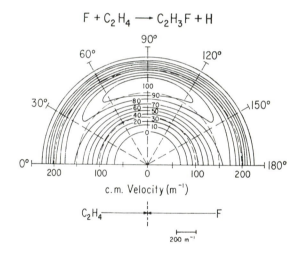

Figure 7.29. Product intensity contour diagram for C_2H_3F produced in the reaction $F + C_2H_4$. (Reproduced from Parson *et al.* 1973.)

sideways peaking can be interpreted in terms of the fluorine atom perpendicularly approaching the plane of C_2H_4. The orbital angular momentum l is very much larger than the rotational angular momentum j of the C_2H_4 molecule and the total angular momentum J, which will be approximately equal to l, will lie in the plane of the ethene molecule. However, in the complex the three heavy atoms will lie in a plane perpendicular to J and the complex will relax to an almost planar geometry in the CCF plane except for the hydrogen atom which is about to leave. The departing hydrogen atom will leave perpendicular to the CCF plane in a direction which is parallel or antiparallel to J. Thus the final velocity v' will be perpendicular to the initial velocity v (figure 7.30).

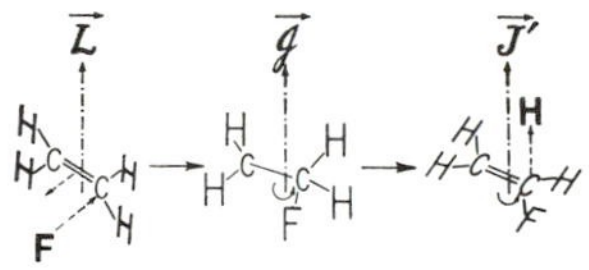

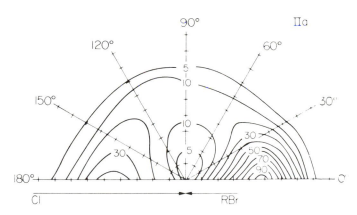

Figure 7.30. Angular momenta for the $F + C_2H_4$ reaction. (Reproduced from Parson *et al.* 1973.)

For some reactions the product intensity diagram shows both forward and backward peaks, with the forward peak being more pronounced. In the reaction

$$Cl + CH_2 = CHCH_2Br \rightarrow ClCH_2CH = CH_2 + Br \qquad (7.38)$$

the forward peak for $ClCH_2CH = CH_2$ is about three times more intense than the backward peak (figure 7.31).

Figure 7.31. Product intensity contour diagram for $ClCH_2CH = CH_2$ produced in the reaction $Cl + CH_2 = CHCH_2Br$. (Reproduced from Herschbach 1976.)

This can be compared with the almost perfect symmetry obtained for $CH_2 = CHCl$ formed in the reaction

$$Cl + CH_2 = CHBr \rightarrow CH_2 = CHCl + Br \qquad (7.39)$$

(figure 7.32) (Cheung *et al.* 1973).

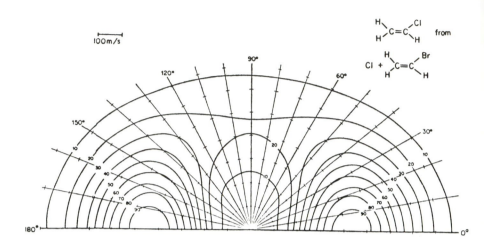

Figure 7.32. Product intensity contour diagram for CH_2=CHCl produced in the reaction $Cl + CH_2$=CHBr. (Reproduced from Herschbach 1973.)

It has been suggested that in the case of reaction 7.38, the complex dissociates as it rotates and has a lifetime of almost half a rotational period (1×10^{-12} s). This has been called an 'osculating' complex by Herschbach.

In some cases a change in mechanism can be observed as the relative translational energy is varied. At an initial relative energy of 3.6 eV the contour diagram (figure 7.23) for the reaction of N^+ with H_2 is asymmetric indicating that at this energy the mechanism is direct (Hansen *et al.* 1980). However, as the energy is lowered below 2 eV, the contour diagram becomes progressively more symmetric. At an energy of 0.98 eV (figure 7.33) it is clearly quite symmetric. Thus it is suggested that at low initial relative energies the reaction proceeds by a complex mechanism.

If a reaction involves a complex, this implies the presence of a well in the potential surface. At higher relative energies one would expect the well to have a smaller influence on the trajectories and the transition to a direct mechanism with increasing energy seems intuitively reasonable. However, a reaction will only proceed by a complex mechanism if the well in the surface is accessible. In the case of NH_2^+, the $^3B_1$ state has an energy of about 6 eV below that of the reactants $N^+(^3P) + H_2$ and one might have expected complex behaviour to persist to higher relative energies. The dynamics of this reaction cannot be interpreted in terms of motion on one potential surface. The $^3B_1$ surface (figure 7.34) has a barrier (of about 3 eV) in the entrance valley and the deep well in this surface is therefore inaccessible to low-energy collisions in which N^+ approaches perpendicularly to the mid-point of the

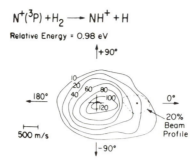

Figure 7.33. Product intensity contour diagram for NH^+ produced in the reaction $N^+ + H_2$ at a relative energy of $0.98\,eV$ showing a symmetrical distribution. (Reproduced from Hansen *et al.* 1980.)

H–H bond. However, two other surfaces, $^3A_2$ and $^3B_2$, correlate with the reactants $N^+(^3P) + H_2$. The $^3B_2$ surface is repulsive and need not be considered further but the $^3A_2$ surface has a shallow well of depth $2.6\,eV$. In the absence of trajectory calculations on this surface one cannot say whether low energy collisions on this surface will give rise to a short-lived complex. It is, however, possible for low-energy collisions to reach the deep well in the $^3B_1$ surface if we consider collision geometries distorted slightly from the perpendicular C_{2v} approach shown in figure 7.34 (Hirst 1978). The symmetry is now C_s and both B_1 and A_2 transform to A'' in the lower symmetry. Thus the intersection between the $^3B_1$ and $^3A_2$ surfaces will be an avoided crossing in C_s symmetry and there is now no barrier between reactants and the lower $^3A''$ surface which correlates with the $^3B_1$ surface. The transition from a complex to a direct mechanism for relative energies of about $1.5\,eV$ can be understood in terms of the increasing probability of non-adiabatic transitions from the lower $^3A''$ to the upper $^3A''$ surfaces as the energy increases.

7.3.4. State Selection of Reactants

7.3.4.1. Vibrational Excitation

In section 6.3.3 we discussed the evidence from trajectory calculations that vibrational energy is more effective than translational energy in promoting reaction on surfaces for which the barrier is in the exit channel. The reaction

$$K + HCl(v = 0) \rightarrow KCl + H \qquad (7.40)$$

is endothermic by $4\,kJ\,mol^{-1}$ and would be expected to have a potential surface with the barrier in the exit channel. The effect of vibrational

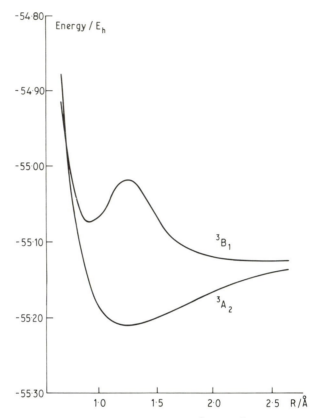

Figure 7.34. Cuts through the entrance channels for the $^3A_2$ and $^3B_1$ potential surfaces of NH_2^+ for C_{2v} geometries with $r_{HH} = 0.74$ Å.

excitation has been demonstrated in a molecular beam experiment by Odiorne *et al.* (1971) in which HCl molecules in the HCl beam were excited to the $v = 1$ level by the radiation from a pulsed HCl chemical laser. They found that the cross-section for reaction with HCl in the $v = 1$ state is approximately 100 times that for ground state HCl. More recent studies have been made in which particular rotational levels of the $v = 1$ state of HCl have been selected. Significant decreases in reactivity were observed as J increased from 1 to 4. This may due to the need for preferential orientation of the reactant molecule before reaction.

The reaction

$$Sr + HF \rightarrow SrF + H \qquad (7.41)$$

is endothermic by 27 kJ mol^{-1} for HF in the ground vibrational state. Karny and Zare (1978) have shown that vibrational excitation is much more effective than translational excitation in promoting reaction. The reaction cross-section for $HF(v = 1)$, for an initial relative translational energy of 7 kJ mol^{-1}, is about 15 times that for $HF(v = 0)$ at the higher translational energy of 54 kJ mol^{-1}.

7.3.4.2. Orientational Effects

In our discussion of the reaction of $K + CH_3I$ in section 7.3.2.2 we implied that the potassium attacks the CH_3I preferentially from the iodine end, approaching along the I–C axis. Steric effects can be demonstrated using oriented molecules. In section 7.1.1.1 we outlined the use of a hexapole electrostatic field to focus a beam of spherical top molecules for which the combination of quantum numbers $KM_J/J(J + 1)$ has a particular negative value. This is equivalent to selecting a particular value for the average of $\cos \theta$ where θ is the angle of orientation of the dipole moment with respect to the electric field. It is not possible to produce beams of perfectly oriented molecules but it is possible to restrict $\cos \theta$ to the range 0 to -1. The beam of molecules then passes into a homogeneous electric field to produce a preferential alignment of the molecules with respect to a laboratory coordinate. The alignment of the molecules can be reversed by changing the polarity of this field. The reactive cross-section for the reaction of K and Rb with CH_3I shows a large change when the polarity is reversed. From the direction of the field it is deduced that the larger cross-section is for the case where attack occurs at the iodine end of the molecule. In the reaction of K with CF_3I (Brooks 1973), reaction occurs effectively for both alignments but attack at the CF_3 end is preferred. The product KI is scattered forwards and this is interpreted in terms of collisions with large impact parameters being effective in reaction at the CF_3 end. The K atom flies past the CF_3 group and then extracts an iodine atom to give KI, which is scattered forwards. Reaction at the I end of CF_3I will occur for small impact parameters and the product KI is scattered backwards. One can understand these processes in terms of an electron jump model in which an electron is transferred from K to CF_3I. The CI bond breaks and KI is scattered preferentially in the direction in which the CI bond is pointing. If K is approaching CF_3I along the CI axis, then KI is scattered backwards but if the orientation of CF_3I is reversed, the CI bond is oriented in the forward direction.

Orientational effects can also be investigated by laser excitation of the reactant. Radiation from a pulsed HF laser is linearly polarized and it is possible to select a plane of polarization. If the radiation is used to excite HF molecules, in a collision cell, to the $v = 1$ state, those molecules for which the dipole moment is collinear with the electric vector are preferentially excited. The preferred orientation persists even though the molecule rotates many times before undergoing reaction. In an investigation of the reaction of Sr with HF($v = 1$), Karny *et al.* (1978) were able to distinguish between Sr approaching HF collinearly or broadside. The broadside approach results in the population of higher vibrational levels in SrF.

7.3.5. *Product State Analysis by Laser-Induced Fluorescence*

For reactions in which it is possible to use laser-induced fluorescence to determine the internal states of the product molecule, one can obtain some very detailed information about the reaction dynamics which cannot be deduced from the angular and velocity distributions measured in conventional molecular beam studies. Kinsey (1977) has reviewed the results of laser-induced fluorescence studies in molecular beam work. It is also possible to combine product state analysis and vibrational excitation of the reagent molecule. In the reaction

$$Ba + HF(v = 0) \rightarrow BaF + H \tag{7.42}$$

at a relative energy of $6.7\,\mathrm{kJ\,mol^{-1}}$, the BaF has a maximum population in $v' = 0$ with a rather smaller population in $v' = 1$ and successively decreasing populations in $v' = 2\text{–}5$. If the HF is excited to $v = 1$, the vibrational distribution of the BaF is completely different. Levels from $v' = 1\text{–}12$ are populated with the maximum at $v' = 6$. The different effects of translational and vibrational energy can be seen by comparing these results with the vibrational population obtained for reaction with HF($v = 0$) at a higher initial relative kinetic energy ($43\,\mathrm{kJ\,mol^{-1}}$). The maximum population of BaF is still in the $v' = 0$ level but levels up to $v' = 10$ are now populated (Pruett and Zare 1976). Trajectory calculations suggest that excess translational energy in the reactants appears as translational energy in the products. In this case 79% of the additional translational energy is converted into translational and rotational energy of the products.

In a laser-induced fluorescence study of the reaction

$$H + NO_2 \rightarrow OH + NO \tag{7.43}$$

made in a crossed-beam apparatus (Silver *et al.* 1976), it was possible to determine full rotational state distributions for the $v = 0$ and $v = 1$ states of OH. The rotational levels of OH are widely spaced and the rotational transitions can thus be resolved in the laser-induced fluorescence. It was estimated that about 40% of the available energy appeared as vibrational energy and 20% as rotational energy. Earlier a conventional molecular beam experiment (Haberland *et al.* 1974) had shown that OH is scattered forward and that about 24% of the available energy is converted into translational energy. It was therefore suggested that the reaction occurs on a highly attractive surface.

7.4. Infra-Red Chemiluminescence Studies of Chemical Reactions

For some reactions much detailed information about the distribution of product molecules among internal states can be obtained by observing infrared chemiluminescence. This is not a single collision technique so it is difficult to ensure that the distribution derived from the recorded spectrum corresponds to the nascent distribution. In many experimental procedures rotational relaxation occurs so one can only obtain information about the vibrational distribution. Rotational relaxation is minimized in the arrested relaxation technique developed by Polanyi and Woodall (1972) in which two uncollimated beams of reagent mix in the centre of a vessel in which the background pressure is of the order of 10^{-3} to 10^{-5} torr (figure 7.35).

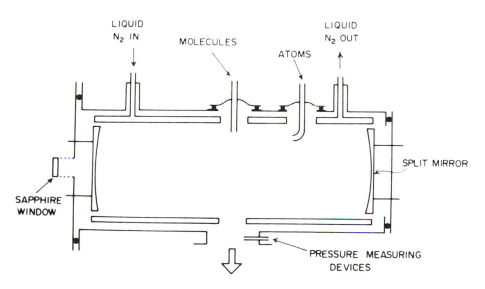

Figure 7.35. Schematic diagram of an apparatus for the measurement of infra-red chemiluminescence by the arrested relaxation technique. (Reproduced from Maylotte *et al.*, *J. Chem. Phys.*, **57**, 1547 (1972).)

Reaction occurs in the region in which the beams mix and the products are rapidly removed from the reaction zone by being pumped away through a large orifice or by condensation on the walls which are maintained at liquid nitrogen temperature. Mirrors at the ends of the vessel collect as much of the radiation as possible which is analysed by a grating spectrometer or interferometer. The spectrum is recorded, and the area under each peak is measured and corrected for any instrumental discrimination effects. Relative populations $N_{v'J'}$ of the vibration–rotation states are then derived by using the expression

$$N_{v'J'} \propto \frac{g_{J'}I(v'J' \to v''J'')}{v^4|R_{v'v''}|^2 S_{J'}F(v'J',v''J'')}, \tag{7.44}$$

where $g_{J'}$ is the rotational degeneracy of level J', I is the intensity of the emission, v is the transition frequency, $S = J'$ for lines in a P-branch and $J' + 1$ for the R-branch, $R_{v'v''}$ is the transition moment for the vibrational transition and F is a vibration–rotation function. The technique cannot give direct information about the populations of states with $v' = 0$ nor can it be applied to homonuclear diatomic molecules or to polyatomic molecules for which the excited vibration is inactive in the infra-red. Many infra-red chemiluminescence studies reported in the literature are for reactions in which the product molecule is a hydrogen halide, frequently HF. The vibrational transitions in HF are strong and the bands are easily resolved. Chemiluminescent reactions have been reviewed by Carrington and Polanyi (1972) and a comprehensive review by Whitehead (1983) on the distribution of energy in product molecules contains details of the results of many investigations of chemiluminescence.

The analysis of infra-red chemiluminescence data from the arrested relaxation technique yields relative detailed rate constants $k(v', J' : T)$ for the formation of the product in a vibration–rotation state $J'v'$. These data can be presented in tabular form but a graphical representation may be easier to interpret. One way of doing this is shown in figure 7.36 in which a relative detailed rate constant is represented by a vertical bar as a function of vibrational level and of rotational level. The diagram also indicates the fractions of the available energy $f_{R'}$ and $f_{V'}$ converted into rotational and vibrational energy respectively. An alternative presentation is in the form of a triangular contour plot as shown in figure 7.37. The two axes at right angles represent vibrational V' and rotational energies R' and the contours join values of V' and R' which have the same detailed rate constant. The contour plot is constrained to lie within a triangle because

$$f_{R'} + f_{V'} + f_{T'} = 1, \tag{7.45}$$

where $f_{T'}$ is the fraction of available energy converted into translational energy. The third side of the triangle joins the points $(0, V'_{max})$ and $(R'_{max}, 0)$ where V'_{max} and R'_{max} are equal to the total available energy. The fraction of energy going into translational energy is represented by the distance from the origin to the hypotenuse of the triangle. Vibrational energies corresponding to particular vibrational levels are also indicated in the diagram.

Figure 7.37 represents data for the reaction

$$F + H_2 \to FH + H \tag{7.46}$$

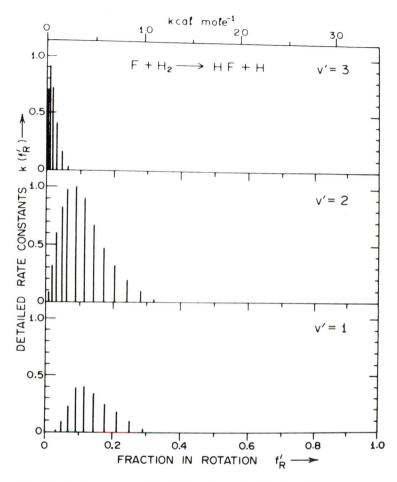

Figure 7.36. Detailed rate constants $k(f_{R'})$ for the reaction $F + H_2$. (Reproduced from Polanyi and Woodall 1972.)

(Polanyi and Woodall 1972) and it is clear that the maximum detailed rate constant $k(v', J' : T)$ is for the formation of HF with $v' = 2$ and that the amount of rotational excitation is low. The fractions of the available energy $\langle f_V \rangle$, $\langle f_R \rangle$ and $\langle f_T \rangle$ converted into vibrational, rotational and translational energy are 0.66, 0.08 and 0.26, respectively. The potential surface for the reaction is repulsive and the substantial vibrational excitation is a result of mixed energy release. For reactions on repulsive surfaces, rotational excitation is thought to occur by the release of energy on product separation. The low rotational excitation in the reaction can be interpreted in terms of a near collinear transition state and thus the repulsive energy release can produce only a small torque on the product molecule.

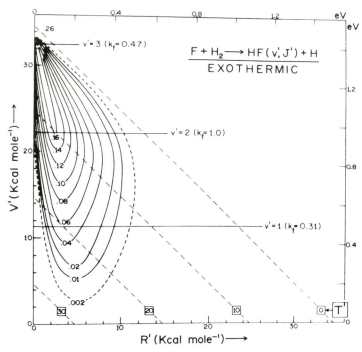

Figure 7.37. Triangular plot of relative rate constants $k(v', J': T)$ plotted as functions of V' and R', the vibrational and rotational energies, for the reaction $F + H_2 \rightarrow HF + H$. (Reproduced from Polanyi and Woodall 1972.)

The infra-red chemiluminescence technique has also been applied to the reactions of H with diatomic halogen molecules. In the reaction of H with F_2, 58% of the available energy is converted into vibrational energy of HF with the maximum population being for $v' = 6$. Only 3% of the energy is channelled into rotation. The potential energy surface for this reaction is repulsive and because of the light atom anomaly, one expects mixed energy release to be unimportant. The values of $\langle f_V \rangle$ for the reactions $H + Cl_2$ and $H + Br_2$ are 0.39 and 0.68, respectively. The variation in going from Cl_2 to Br_2 can be interpreted in terms of the potential surface becoming more attractive. The barrier height decreases from $7.5 \, kJ \, mol^{-1}$ to $3.8 \, kJ \, mol^{-1}$ in going from $H + Cl_2$ to $H + Br_2$ and the barrier is expected to move to an earlier location in the entrance channel. Thus the fraction of attractive energy release would be expected to increase. For $H + F_2$ the barrier height is $10 \, kJ \, mol^{-1}$ and thus the observed vibrational energy release seems anomalous. Trajectory calculations suggest that the anomaly for F_2 is not due to the effect of changing mass combinations. The larger attractive energy release for the $H + F_2$ reaction can be attributed to the short range of the F–F repulsive interaction.

Suggestions for Further Reading

R. B. Bernstein (1982) *Chemical Dynamics via Molecular Beam and Laser Techniques,* Clarendon Press, Oxford.

P. N. Clough and J. Geddes (1981) *J. Phys. E,* **14**, 519.

J. Geddes (1982) *Contemp. Physics,* **23**, 233.

R. Grice (1975) in *Molecular Scattering: Physical and Chemical Applications,* ed. K. P. Lawley, p. 247, Wiley, London.

R. Grice (1979) *Farad. Disc. Chem. Soc.,* **67**, 16.

P. R. Brooks and E. F. Hayes (1977) *State-to-State Chemistry,* American Chemical Society, Washington D.C.

D. R. Herschbach (1973) *Farad. Disc. Chem. Soc.,* **55**, 233.

D. R. Herschbach (1976) *Pure and Applied Chemistry,* **47**, 61.

J. L. Kinsey (1972) in *M. T. P. International Review of Science, Physical Chemistry, Series 1,* Vol. 9, ed. J. C. Polanyi, p. 173, Butterworths, London.

R. D. Levine and R. B. Bernstein (1974) *Molecular Reaction Dynamics,* Clarendon Press, Oxford.

I. W. M. Smith (1980) *Kinetics and Dynamics of Elementary Gas Reactions,* Butterworths, London.

References

R. B. Bernstein (1982) *Chemical Dynamics via Molecular Beam and Laser Techniques,* Clarendon Press, Oxford.

P. R. Brooks (1973) *Farad. Disc. Chem. Soc.,* **55**, 299.

P. R. Brooks (1976) *Science,* **193**, 11.

U. Buck (1975) in *Molecular Scattering: Physical and Chemical Applications,* ed. K. P. Lawley, p. 313, Wiley, London.

U. Buck (1974) *Rev. Mod. Phys.,* **46**, 369.

U. Buck and H. Pauly (1971) *J. Chem. Phys.,* **54**, 1929.

T. Carrington and J. C. Polanyi (1972) in *M. T. P. International Review of Science, Physical Chemistry, Series 1,* Vol. 9, ed. J. C. Polanyi, p. 135, Butterworths, London.

J. T. Cheung, J. D. McDonald and D. R. Herschbach (1973) *J. Amer. Chem. Soc.,* **95**, 7889.

H. W. Cruse, P. J. Dagdigian and R. N. Zare (1973) *Farad. Disc. Chem. Soc.,* **55**, 277.

M. A. D. Fluendy and K. P. Lawley (1973) *Chemical Applications of Molecular Beam Scattering,* Chapman and Hall, London.

W. R. Gentry (1979) in *Gas Phase Ion Chemistry,* Vol. 2, ed. M. T. Bowers, p. 221, Academic Press, New York.

K. T. Gillen, A. M. Rulis and R. B. Bernstein (1971) *J. Chem. Phys.,* **54**, 2831.

R. Grice (1981a) in *Gas Kinetics and Energy Transfer,* Vol. 4, ed. P. G. Ashmore and R. J. Donovan, p. 1, Royal Society of Chemistry, London.

R. Grice (1981b) *Acc. Chem. Research,* **14**, 37.

H. Haberland, P. Rohmer and K. S. Schmidt (1974) *Chem. Phys.,* **5**, 298.

S. G. Hansen, J. M. Farrar and B. H. Mahan (1980) *J. Chem. Phys.*, **73**, 3750.

Z. Herman and R. Wolfgang (1972) in *Ion–Molecule Reactions*, Vol. 2, ed. J. L. Franklin, p. 553, Butterworths, London.

D. M. Hirst (1978) *Mol. Phys.*, **35**, 1559.

Z. Karny and R. N. Zare (1978) *J. Chem. Phys.*, **68**, 3360.

Z. Karny, R. C. Estler and R. N. Zare (1978) *J. Chem. Phys.*, **69**, 5199.

J. L. Kinsey (1977) *Ann. Rev. Phys. Chem.*, **2**, 349.

W. S. Koski (1975) in *Molecular Scattering: Physical and Chemical Applications*, ed. K. P. Lawley, p. 185, Wiley, London.

J. D. McDonald, P. R LeBreton, Y. T. Lee and D. R. Herschbach (1972) *J. Chem. Phys.*, **56**, 769.

G. C. Maitland, M. Rigby, E. B. Smith and W. A Wakeham (1981) *Intermolecular Forces*, Clarendon Press, Oxford.

E. A. Mason, R. J. Munn and F. J. Smith (1971) *Endeavour*, **30**, 91.

M. Menzinger (1980) in *Potential Energy Surfaces* ed. K. P. Lawley, p. 1, Wiley, Chichester.

W. B. Miller, S. A. Safron and D. R. Herschbach (1967) *Disc. Farad. Soc.*, **44**, 108.

C. V. Nowikow and R. Grice (1979) *J. Phys. E.*, **12**, 515.

Ch. Ottinger (1984) in *Gas Phase Ion Chemistry*, Vol. 3, ed. M. T. Bowers, p.250, Academic Press, New York.

T. J. Odiorne, P. R. Brooks and J. V. V. Kasper (1971) *J. Chem. Phys.*, **55**, 1980.

J. M. Parson, K. Shobotake, Y. T. Lee and S. A. Rice (1973) *Farad. Disc. Chem. Soc.*, **55**, 344.

J. C. Polanyi and K. B. Woodall (1972) *J. Chem. Phys.*, **57**, 1574.

J. G. Pruett and R. N. Zare (1976) *J. Chem. Phys.*, **64**, 1774.

S. J. Riley and D. R. Herschbach (1973) *J. Chem. Phys*, **58**, 27.

A. M. Rulis and R. B. Bernstein (1972) *J. Chem. Phys.*, **57**, 5497.

T. P. Schafer, P. E. Siska, J. M. Parson, F. P. Tully, Y. C. Wong and Y. T. Lee (1970) *J. Chem. Phys.*, **53**, 3385.

J. A. Silver, W. L. Dimpfl, J. H. Brophy and J. L. Kinsey (1976) *J. Chem. Phys.*, **65**, 1811.

P. E. Siska (1973) *J. Chem. Phys.*, **59**, 6052.

J. P. Toennies (1974) in *Physical Chemistry, An Advanced Treatise*, Vol VIA, ed. W. Jost, p. 228, Academic Press, New York.

T. T. Warnock and R. B. Bernstein (1968) *J. Chem. Phys.*, **49**, 1878.

J. C. Whitehead (1983) in *Comprehensive Chemical Kinetics*, Vol. 24, ed. C. H. Bamford and C. F. H. Tipper, p. 357, Elsevier, Amsterdam.

R. Wolfgang and J. R. Cross (1969) *J. Phys. Chem.*, **73**, 743.

R. N. Zare and P. J. Dagdigian (1974) *Science*, **185**, 739.

R. N. Zare (1979) *Farad. Disc. Chem. Soc.*, **67**, 7.

intermolecular potentials and van der Waals molecules

8.1. Intermolecular Forces

In section 7.3 we discussed the determination of intermolecular potentials from molecular beam studies of elastic scattering. We suggested that in the case of two atoms, the interatomic potential is qualitatively similar to that for a bound diatomic molecule. However, the depth of the well is much smaller than the dissociation energy for a typical diatomic molecule and the minimum occurs at a much larger internuclear separation. In this chapter we consider van der Waals molecules in which two species A and B, which may be atoms or molecules, are held together by intermolecular forces. There has been increasing interest over the past 15 years in the structure, bonding and potential energy surfaces for these weakly bound molecules. In this section we will discuss briefly the nature of intermolecular forces and in subsequent sections we will consider experimental techniques for obtaining structural information about van der Waals molecules and how we can deduce the potential energy surface from the structural information. For a fuller discussion of the nature of intermolecular potential functions the reader is referred the recent comprehensive monograph by Maitland *et al.* (1981). It is usual to discuss long-range interactions and short-range interactions separately because quite different methods are used for calculating these interactions.

8.1.1. Long-Range Interactions

If the two species are separated by a large distance, it is reasonable to assume that there is no overlap between the electrons of one species with those of the other. The interaction between the two molecules can then be treated by perturbation theory and it is not necessary to consider the exchange of electrons between the two species (Pople 1982). Three types of interaction are usually considered.

If the two molecules A and B have permanent dipole moments $\boldsymbol{\mu}_A$ and $\boldsymbol{\mu}_B$, there will be a classical electrostatic interaction between the two dipoles separated by a distance R

$$U_{el} = \frac{1}{4\pi\epsilon_0} \frac{1}{R^3} [\boldsymbol{\mu}_A \cdot \boldsymbol{\mu}_B - 3(\boldsymbol{\mu}_A \cdot \hat{\boldsymbol{R}})(\hat{\boldsymbol{R}} \cdot \boldsymbol{\mu}_B)], \tag{8.1}$$

where $\hat{\boldsymbol{R}}$ is a unit vector defining the direction of the centre of dipole B with respect to that of dipole A. If all relative orientations were equally probable, the dipole–dipole interaction would average to zero. However each relative orientation must be weighted by a Boltzmann factor $\exp[-U(\boldsymbol{\omega}_A, \boldsymbol{\omega}_B)/kT]$, where $\boldsymbol{\omega}_A$ and $\boldsymbol{\omega}_B$ represent the orientations of $\boldsymbol{\mu}_A$ and $\boldsymbol{\mu}_B$. On calculating the average over all orientations

$$\langle U_{el} \rangle = \frac{\iint U_{el}(\boldsymbol{\omega}_A, \boldsymbol{\omega}_B)\exp[-U(\boldsymbol{\omega}_A, \boldsymbol{\omega}_B)/kT]\,d\boldsymbol{\omega}_A\,d\boldsymbol{\omega}_B}{\iint \exp[-U(\boldsymbol{\omega}_A, \boldsymbol{\omega}_B)/kT]\,d\boldsymbol{\omega}_A\,d\boldsymbol{\omega}_B}, \tag{8.2}$$

assuming that the intermolecular potential is small compared with kT, one obtains the following expression for the interaction potential $\langle U_{el} \rangle$ between two dipoles of magnitudes μ_A and μ_B.

$$\langle U_{el} \rangle = -\frac{2}{3} \frac{\mu_A^2 \mu_B^2}{R^6 kT(4\pi\epsilon_0)^2} . \tag{8.3}$$

In addition to this direct electrostatic interaction, the dipole moment on molecule A will induce a dipole moment in species B (irrespective of whether B is polar or non-polar) given by

$$\boldsymbol{\mu}_{IND} = \alpha_B \boldsymbol{E}, \tag{8.4}$$

where $\boldsymbol{E}$ is the electric field at B due to the dipole $\boldsymbol{\mu}_A$ and α_B is the isotropic polarizability of species B. The interaction between the permanent dipole on A and the induced dipole on B results in an attractive interaction potential given by

$$U_{IND} = -\tfrac{1}{2}\alpha_B |\boldsymbol{E}|^2. \tag{8.5}$$

An expression for $\boldsymbol{E}$ can be developed using classical electrostatics and assuming that both species are polar, on averaging over all orientations, one obtains the following expression for the dipole–induced dipole interaction

$$\langle U_{IND} \rangle = -\frac{1}{(4\pi\epsilon_0)^2 R^6}(\mu_A^2 \alpha_B + \mu_B^2 \alpha_A). \tag{8.6}$$

If just one molecule, A, is polar this reduces to

$$\langle U_{\text{IND}} \rangle = -\frac{1}{(4\pi\epsilon_0)^2 R^6} \mu_A^2 \alpha_B. \tag{8.7}$$

There is a third type of intermolecular interaction which operates even though both species are non-polar. This is the London dispersion force which cannot be understood in terms of classical electrostatics. Let us consider, in the first instance, the interaction between two atoms. On average each atom has a spherically symmetrical charge distribution and is thus non-polar. However, the electronic distribution is fluctuating and at a particular instant in time species A may have an instantaneous dipole. This instantaneous dipole can induce a dipole moment in species B and there will be an attractive potential between the two species. A rigorous discussion of the dispersion interaction requires a quantum mechanical treatment in which the interaction is taken into account by second order perturbation theory and is beyond the scope of this book. The final expression for the potential, which also depends on R^{-6}, is in terms of a double summation over all electronic states of species A and B and is therefore very difficult to calculate exactly. The dispersion interaction is usually evaluated semi-empirically using polarizability and refractive index data. London developed the following approximate formula for the dispersion potential

$$U_{\text{DISP}} = -\frac{3}{2} \frac{1}{R^6 (4\pi\epsilon_0)^2} \left(\frac{I_A I_B}{I_A + I_B} \right) \alpha_A \alpha_B \tag{8.8}$$

in terms of the polarizabilities α_A, α_B and of the ionization energies I_A, I_B of the two species.

Thus in the first approximation, in which only dipole effects are considered, the intermolecular potential can be written in the form

$$U = -\frac{1}{R^6} (A + B + C), \tag{8.9}$$

where A, B and C represent the dipole–dipole, the dipole–induced dipole and dispersion interactions respectively. The dispersion contribution makes the largest contribution in the majority of cases and it is, of course, the only interaction when species A and B are both non-polar.

If dipole–quadrupole and quadrupole–quadrupole interactions are included, they give rise to additional attractive terms which depend on R^{-8} and R^{-10} respectively.

It is also possible to calculate the long-range part of the potential by the methods of quantum chemistry but care must be taken to choose an appropriate model. The method should be capable of yielding the correct asymptotic form for the energy. The wavefunction must be fully antisymmetric with respect to the interchange of electrons and should account for exchange and overlap effects and closed shell repulsions at short internuclear separations. For closed shell systems the SCF method gives a reasonable description of the interaction between permanent multipoles and of induction effects but the description of dispersion effects requires a careful treatment of electron correlation (Kutzelnigg 1977).

8.1.2. *Short-Range Interactions*

At shorter interatomic distances there is overlap between the electron clouds on the two species and electron exchange can no longer be neglected. It is not possible to derive formulae analogous to those in equations 8.3, 8.6 and 8.8 in this region of the potential. The intermolecular potential can be calculated by the methods of quantum chemistry discussed in chapter 5. The configuration interaction, MC–SCF and valence bond methods have been used for systems held together by intermolecular forces. One has to be careful to include the configurations required to give a satisfactory description of the dispersion forces. The intermolecular potential is given by

$$U = E_{AB} - E_A - E_B,   \tag{8.10}$$

where E_{AB}, E_A and E_B are the energies calculated for AB, A and B, respectively. In order to avoid spurious results for the intermolecular potential it is essential to ensure that E_{AB}, E_A and E_B are calculated with comparable accuracy. To avoid basis-set superposition errors, E_{AB}, E_A and E_B should be calculated with the same basis set. The calculation of intermolecular forces has been reviewed by Certain and Bruch (1972) and by Stamper (1975).

8.1.3. *Model Potential Functions*

In view of the difficulty in calculating intermolecular potentials it is not surprising that many approximate analytical forms have been put forward. We mention briefly a few of these here but restrict ourselves to the isotropic potential which simply depends on the distance between the two species. Perhaps the most widely used function is the Lennard-Jones $n-6$ potential which can be written in terms of the well depth ϵ and R_m, the value of R at the potential minimum.

$$U(R) = \epsilon \left[\frac{6}{n-6} \left(\frac{R_m}{R} \right)^n - \frac{n}{n-6} \left(\frac{R_m}{R} \right)^6 \right].   \tag{8.11}$$

This clearly has the correct R^{-6} dependence for large R. The parameter n is often given the value 12 and thus

$$U(R) = \epsilon \left[\left(\frac{R_m}{R} \right)^{12} - 2 \left(\frac{R_m}{R} \right)^6 \right],   \tag{8.12}$$

which can be written alternatively as

$$U(R) = 4\epsilon\left[\left(\frac{\sigma}{R}\right)^{12} - \left(\frac{\sigma}{R}\right)^{6}\right], \tag{8.13}$$

where σ is the value of R for which $U(R)$ is zero. A more realistic description of the repulsive part of the potential is given by an exponential term as in the Buckingham potential

$$U(R) = \epsilon\left[\frac{6}{\alpha-6}\exp\left\{-\alpha\left(\frac{R}{R_m}-1\right)\right\} - \frac{\alpha}{\alpha-6}\left(\frac{R_m}{R}\right)^{6}\right]. \tag{8.14}$$

These functions have the advantage of being simple but the repulsive parts have no theoretical basis and the functions are not sufficiently flexible to represent all of the known data for a given system. A function suggested by Barker and Pompe containing 11 disposable parameters in addition to ϵ and R_m is capable of giving a much more accurate representation of the intermolecular potential

$$\frac{U(\bar{R})}{\epsilon} = \exp[\alpha(1-\bar{R})]\sum_{i=0}^{5} A_i(\bar{R}-1)^i - \sum_{j=0}^{2}\frac{C_{2j+6}}{(\delta+\bar{R})^{2j+6}}. \tag{8.15}$$

In this function $\bar{R} = R/R_m$ and α, A_i, C_{2j+6} and δ are adjustable parameters. For further details of analytical potential functions the reader is referred to Maitland *et al.* (1981).

8.2. Diatomic Species

The simplest type of van der Waals molecule is a diatomic species composed of two atoms which are unable to form a chemical bond with each other. Such molecules are present in low concentrations in rare gases. At pressures near atmospheric, the concentration of dimers is of the order of 0.1% of the atomic concentration. The form of the potential curve for a diatomic van der Waals molecule is shown in figure 8.1. The diagram assumes that the molecule is in the $J = 0$ rotational state as will be the case if the atoms approach each other along the line of centres with zero impact parameter b. However if the impact parameter is non-zero, the resulting molecule will have an angular momentum L given by $\mu b v$, where μ is the reduced mass $m_A m_B/(m_A + m_B)$ and v is the relative velocity. The rotational kinetic energy has to be taken into account in the Hamiltonian for the nuclear motion as discussed in sections 2.1 and 2.8.2. The effective potential takes the form

$$U_{\text{eff}}(R) = U(R) + \frac{L^2}{2\mu R^2}, \tag{8.16}$$

where $U(R)$ is the potential energy function that would be obtained by

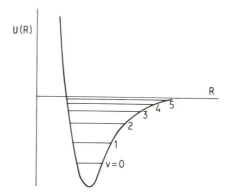

Figure 8.1. A potential curve for a diatomic van der Waals molecule in the rotational state $J = 0$.

solution of the electronic Schrödinger equation. The potential function $U_{eff}(R)$ has a local maximum as shown in figure 8.2 and there may be quasi-bound vibrational levels.

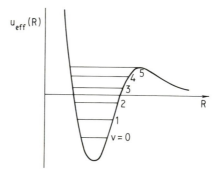

Figure 8.2. A potential curve for a diatomic van der Waals molecule with $J > 0$. The vibrational levels for $v = 3, 4, 5$ are quasi-bound.

In the case where the angular momentum L is zero, in order to form a van der Waals molecule it is necessary for the pair of atoms to collide with a third body in such a way that the final energy is less than that of the separated atoms. If L is greater than zero there are two possibilities. A stable van der Waals molecule can be formed just as in the case where L is zero but it is also possible to form a metastable or quasi-bound van der Waals molecule which has an energy between that of the separated atoms and the energy at the local maximum in $U_{eff}(R)$. Van der Waals molecules in quasi-bound states can dissociate by quantum mechanical tunnelling through the barrier.

Electronic spectra of the species Ne_2, Ar_2, Kr_2 and Xe_2 have been observed in the ultra-violet by Tanaka and coworkers (see Freeman *et al.* 1974) and by Colbourn and Douglas (1976). In order to observe the spectra of the dimers it is necessary to use a path length of several metres. The ultra-violet spectrum of argon shows an intense absorption near 1067 Å due to the atomic transition from the ground state $^1S_0$ to the $^3P_1^0$ state. Bands due to the transition from ground state $Ar_2(^1\Sigma_g^+)$ to an excited state $(^1\Sigma_u^+)$ are observed to the long wavelength side of the atomic absorption line in spectra taken with an argon pressure of 10 torr (1 torr $= 1.33322 \times 10^2 \, N \, m^{-2}$) at 77 K. The bands can be analysed in terms of progressions from ground state vibrational levels. Tanaka and his coworkers were able to make a rotational analysis only for Ne_2 but Colbourn and Douglas (1976) were able to achieve higher resolution for Ar_2 and resolve the rotational structure for one band system.

The spectra can be analysed by the conventional methods of spectroscopy but in deriving information about the potential energy function one must take care not to use models which are inappropriate to van der Waals molecules. For example the expression in equation 2.17 for the vibration–rotation levels of a diatomic molecule is obtained by solving the nuclear Schrödinger equation for a Morse potential. Such a potential function is not capable of giving an adequate representation of an intermolecular potential so the spectra of van der Waals molecules should not be fitted to such energy level expressions. The RKR method can be used for Ar_2 and Ne_2 (where rotational structure has been resolved) to obtain the potential curve up to the highest observed vibration–rotation level. Since the RKR method determines the turning points of the vibrational motion it does not assume any functional form for the potential. From a potential calculated in this way one can obtain R_m, the internuclear distance at the potential minimum. The other characteristic of the potential curve is the well depth, ϵ, which is equivalent to the dissociation energy $\tilde{D}_e$ for an ordinary molecule. Dissociation energies for chemically bound molecules are most commonly obtained from vibrational spectra by the Birge–Sponer extrapolation. For a Morse potential or a harmonic potential to which cubic and quartic terms have been added, the vibrational term values $G(v)$ can be written, to a first approximation, as

$$G(v) = \omega_e(v + \tfrac{1}{2}) - x(v + \tfrac{1}{2})^2 \tag{8.17}$$

(see the discussion in section 2.1.2). Here we have neglected the cubic terms in $(v + \tfrac{1}{2})$ from equation 2.14. The spacing $\Delta G_{v+\frac{1}{2}}$ between successive levels v and $v+1$ is given by

$$\Delta G_{v+\frac{1}{2}} = \omega_e - 2x(v+1) \tag{8.18}$$

and thus a plot of $\Delta G_{v+\frac{1}{2}}$ against v should be a straight line which will cross the v axis at v_{max}, the maximum allowed value of v. In the Birge–Sponer method, the dissociation energy $\tilde{D}_0$ is obtained by extrapolating the plot from the highest observed value v_h of v to the intersection with the v-axis. The area under the line from the highest observed level v_h to v_{max} is added to the energy of the level v_h to yield an estimate of the dissociation energy. Clearly the procedure is valid only if equation 8.17 is a satisfactory representation of the term values of the vibrational levels. In some cases the Birge–Sponer plot shows very distinct curvature. If vibrational levels are observed to high values of v, errors in the extrapolation procedure will be relatively unimportant. However, in van der Waals molecules the potential cannot be represented adequately by a Morse function or by a harmonic potential with cubic and quartic terms. Thus the vibrational levels are not adequately represented by equation 8.17 and use of the Birge–Sponer procedure will lead to an erroneous value for the dissociation energy.

LeRoy and Bernstein (see LeRoy 1973) have proposed a different procedure for extrapolation from the highest observed vibrational level to the dissociation limit. This method can be applied to van der Waals molecules and can also be used for chemically bound molecules to provide a more accurate estimate than that given by the Birge–Sponer method. It is assumed that in the asymptotic region of large R, where the potential energy is approaching the dissociation limit, the potential energy can be written as

$$U(R) = \epsilon - \frac{C_n}{R^n}, \qquad (8.19)$$

where C_n is positive. In the case of van der Waals molecules $n = 6$ but for some states of chemically bound species (e.g. the $B\,^3\Pi_{0u}^+$ state of Cl_2) $n = 5$. Following a semi-classical Wentzel–Kramers–Brillouin approach, as outlined in section 2.3, one can write the quantum number v for a given level, in the absence of rotation, in the form

$$v + \tfrac{1}{2} = \left(\frac{8\mu}{h^2}\right)^{1/2} \int_{R_L}^{R_R} [G(v) - U(R)]^{1/2} \, \mathrm{d}R, \qquad (8.20)$$

where R_L and R_R are the inner and outer turning points for that vibrational level and μ is the reduced mass. An expression for the density of levels can be obtained by differentiation of equation 8.20 to give

$$\frac{\mathrm{d}v}{\mathrm{d}G(v)} = \left(\frac{2\mu}{h^2}\right)^{1/2} \int_{R_L}^{R_R} [G(v) - U(R)]^{-1/2} \, \mathrm{d}R. \qquad (8.21)$$

The major contribution to the integral in equation 8.21 comes from the regions near R_L and R_R where the integrand becomes singular. Because of the anharmonicity of the potential, the contribution from R_R becomes dominant as the dissociation limit is approached and R_L can be set to zero without introducing any significant error. With this approximation, evaluation of the integral gives

$$\frac{dG(v)}{dv} = K_n[\epsilon - G(v)]^{(n+2)/2n}, \tag{8.22}$$

where K_n represents a rather complicated function involving gamma functions. Integration of equation 8.22, provided $n \neq 2$, yields

$$G(v) = \epsilon - [(v_{max} - v)H_n]^{2n/(n-2)}, \tag{8.23}$$

where $H_n = [(n-2)/2n]K_n$. The spacing between adjacent levels $\Delta G_{v+\frac{1}{2}}$ is approximately equal to $dG(v+\frac{1}{2})/dv$ and thus from equation 8.23 one obtains the relationship

$$(\Delta G_{v+\frac{1}{2}})^{(n-2)/(n+2)} = (v_{max} - v - \tfrac{1}{2})\left[\frac{2n}{n-2}H_n^{2n/(n-2)}\right]^{(n-2)/(n+2)}. \tag{8.24}$$

Thus a plot of $(\Delta G_{v+\frac{1}{2}})^{(n-2)/(n+2)}$ against v will be a straight line. This is in contrast to the Birge–Sponer extrapolation in which $\Delta G_{v+\frac{1}{2}}$ is plotted against v. For the case of $n = 6$, v_{max} can be obtained by extrapolating the plot of $(\Delta G_{v+\frac{1}{2}})^{1/2}$ against v. This is illustrated for the case of Ar_2 in figure 8.3 in which a plot of $(\Delta G_{v+\frac{1}{2}})^{1/2}$ against v (line T) is compared with a conventional Birge–Sponer extrapolation (line E).

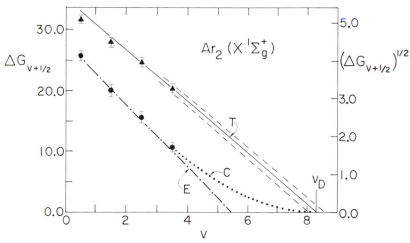

Figure 8.3. Extrapolation of vibrational energy levels of Ar_2. Extrapolations T and C are obtained using equation 8.24 and E is the conventional Birge–Sponer extrapolation. (Reproduced from LeRoy, *J. Chem. Phys.*, **57**, 573 (1972).)

Equation 8.23 and equations derived from it can also be used to determine n and C_n. These applications are discussed in detail by LeRoy (1973). More recently LeRoy and Lam (1980) have extended the method so that one can obtain improved dissociation energies for diatomic molecules for which there are no vibrational data for levels near to the dissociation limit.

The potential function for Ar_2 derived from the spectroscopic data by the RKR method and the use of the method outlined above for the determination of the dissociation energy is in good agreement with those derived from scattering experiments and from bulk properties of gaseous, liquid and solid argon.

The intermolecular potential between two helium atoms is very small and there are no bound vibrational states for He_2. For other rare gas dimers well depths vary from $0.2\,kJ\,mol^{-1}$ for Ne_2 to $2.2\,kJ\,mol^{-1}$ for Xe_2. The potential curve for Ar_2, with $L = 0$, has 8 vibrational levels whereas Xe_2 is predicted to have 25 or 26 bound vibrational levels. When non-zero values of L are taken into account many more levels are possible and for Ar_2 there are about 170 levels.

More recently, spectra have been observed for NaAr, NaNe and XeF (Levy 1981) in supersonic beams. In the case of NaAr, the fluorescence excitation spectrum and the dispersed fluorescence spectrum have been observed. These spectra give information about the well in the ground state $X^2\Sigma^+$ potential and the portion of the repulsive wall up to a few thousand cm^{-1} above the dissociation limit. The potential curve for the excited $A^2\Pi$ state resembles that of a bound molecule. The well depth is of the order of $560\,cm^{-1}$ which is much deeper than that of the ground state ($40\,cm^{-1}$) and the value of R_m is $2.91\,Å$ compared with $4.99\,Å$ for the ground state. Consequently the spectra only give information about excited vibrational levels with $v' = 7$–11.

8.3. Triatomic van der Waals Molecules

8.3.1. Anisotropic Potentials and the Hamiltonian Operator

Triatomic van der Waals molecules are typically formed from an inert gas atom and a diatomic molecule. Many such molecules have been observed by a variety of techniques (Howard 1975, Levy 1981) but so far it has been possible to obtain detailed potential surfaces for only a limited number of species. In subsequent sections we shall discuss two examples, namely Ar–H_2 and Ar–HCl. It is convenient to discuss the potential surface in terms of the coordinate system (R, r, θ) shown in figure 8.4. R is the distance between the inert gas atom A and the centre of mass of the diatomic molecule XY. The potential is no longer simply a function of R but will be anisotropic. The van der Waals interaction is very weak and it is therefore reasonable to separate the van der Waals potential $V(R, r, \theta)$ from the potential $V_{XY}(r)$ for the molecule XY thus:

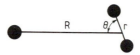

Figure 8.4. Coordinate system for a van der Waals molecule.

$$V_{\text{tot}}(R, r, \theta) = V_{\text{XY}}(r) + V(R, r, \theta). \tag{8.25}$$

The van der Waals potential can be written as an expansion in terms of Legendre polynomials $P_k(\cos \theta)$

$$V(R, r, \theta) = \sum_k V_k(R, r) P_k(\cos \theta). \tag{8.26}$$

This is not very satisfactory for very small values of R because many terms would have to be included but for values of R appropriate to bound van der Waals molecules, the expansion is rapidly convergent. If the diatomic molecule XY is homonuclear, the expansion in equation 8.26 is restricted to even values of k. The van der Waals potential depends only weakly on the internuclear distance r of the diatomic molecule XY and it can be taken into account by the following expansion

$$V(R, r, \theta) = V(R, \xi, \theta) = \sum_k \sum_\lambda \xi^\lambda P_k(\cos \theta) V_{k\lambda}(R), \tag{8.27}$$

where ξ can be written as $(r - r_0)/r_0$ in terms of an isotope-independent constant r_0. If one is only concerned with the ground vibrational state of a van der Waals molecule one can use a potential which is independent of r. This can be written

$$V(R, \theta) = \sum_k V_k(R) P_k(\cos \theta). \tag{8.28}$$

The Hamiltonian for the nuclear motion of a van der Waals molecule can be written in terms of space-fixed coordinates or in terms of molecule-fixed coordinates. A Hamiltonian expressed in terms of space-fixed axes is appropriate for molecules such as Ar–H_2 for which the coupling is weak so that one can consider the diatomic XY to be vibrating and rotating freely within the complex which is itself vibrating and rotating. This would be the case if the anisotropic terms in equation 8.27 were all zero. The Hamiltonian in terms of space-fixed axes has the form

$$H = \frac{P_R^2}{2\mu_{\text{A,XY}}} + \frac{l^2}{2\mu_{\text{A,XY}} R^2} + V(R, r, \theta) + H_{\text{XY}}(r), \tag{8.29}$$

where P_R is the vibrational momentum in the coordinate R, $(h/2\pi)l$ is the operator for the rotational angular momentum of the AXY molecule and H_{XY} is the vibration–rotation Hamiltonian for the isolated diatomic molecule

XY. $H_{XY}(r)$ will have the form

$$H_{XY} = \frac{P_r^2}{2\mu_{XY}} + \frac{j^2}{2\mu_{XY}r^2} + V_{XY}(r), \tag{8.30}$$

where P_r is the vibrational momentum in coordinate r, $(h/2\pi)j$ is the operator for the rotational angular momentum of XY and $V_{XY}(r)$ is the potential energy function for XY. In this Hamiltonian all distances and operators are expressed in terms of space-fixed axes. The Hamiltonian in terms of molecule-fixed axes differs only in that the distances and operators are expressed with respect to molecule-fixed axes and that the angular momentum operator l for the rotation of the AXY molecule is replaced by $J-j$, where J is the operator for the total angular momentum. Thus

$$H = \frac{P_R^2}{2\mu_{A,XY}} + \frac{(J-j)^2}{2\mu_{A,XY}R^2} + V(R, r, \theta) + H_{XY}(r). \tag{8.31}$$

The quantum numbers j and l associated with the operators j and l are not strictly good quantum numbers and one should use the quantum number J for the total rotational angular momentum. J can take the range of values

$$J = j+l, j+l-1, \ldots, |j-l|. \tag{8.32}$$

Van der Waals molecules can be classified as weak coupling, strong coupling or semi-rigid according to the relative values of the average effective anisotropy $\overline{\Delta V_0}$, the spacing $\Delta E_{XY}(j)$ between rotational levels in the diatomic and the spacing $\Delta E_{AXY}(l)$ between the rotational levels of the complex. The average effective anisotropy $\overline{\Delta V_0}$ is defined as the sum of all the $V_{\lambda k}(R)$ terms (except for $k = 0$) in equation 8.27 averaged over the ranges of R and ξ appropriate for the complex. In the weak coupling case (such as Ar–H$_2$), $\Delta E_{AXY}(l) \lesssim \overline{\Delta V_0} \ll \Delta E_{XY}(j)$ and j is a good quantum number. In the strong coupling case (such as ArHCl) the potential is more anisotropic, $\Delta E_{AXY}(l) < \Delta E_{XY}(j) \lesssim \overline{\Delta V_0}$, and XY is no longer able to perform virtually free rotation. In this case there is mixing between different j levels and the Hamiltonian expressed in terms of molecule-fixed axes is more appropriate. If the potential is extremely anisotropic (as in the case of KrClF), $\Delta E_{AXY}(l)$, $\Delta E_{XY}(j) \ll \overline{\Delta V_0}$, and the complex can be treated as an ordinary molecule. Different approaches are used for the determination of the potential function for the weak and strong coupling cases and we will discuss Ar–H$_2$ and ArHCl separately in subsequent sections.

It is not possible to calculate directly the potential surface for a triatomic van der Waals molecule by inverting the spectroscopic data. One has to proceed by taking a potential function, of the form of equations 8.27 or 8.28, which contains several adjustable parameters and calculating from it a theoretical spectrum. The variable parameters are successively adjusted until the best fit with the experimental spectrum is obtained. A detailed discussion of the methods used for the calculation of the spectrum from the potential surface is beyond the scope of this book. A comprehensive survey has been given by LeRoy and Carley (1980) and we restrict ourselves to a brief outline. One approach calculates the eigenvalues and eigenfunctions of the Hamiltonian from the set of coupled equations used in quantum scattering theory. These equations are the same as those used to describe the elastic and inelastic scattering of the atom A by the diatomic molecule XY. The equations are solved either by numerical integration for a series of trial energies until all of the boundary and continuity requirements are satisfied or by expansion of the wavefunction in terms of a set of basis functions and solving the resulting secular equations. An alternative approach, which is less expensive computationally, assumes that the vibrational–rotational motion of the diatomic molecule XY is much more rapid than the stretching of the van der Waals bond. By analogy with the Born–Oppenheimer separation, the wavefunction for bending and stretching $\psi_{bs}(\theta, R)$ is written as a product of a bending wavefunction $\phi_b(\theta; R)$ which depends parametrically on R, and a stretching wavefunction $\chi_{bs}(R)$. The subscripts b and s represent bending and stretching quantum numbers respectively.

$$\psi_{bs}(\theta, R) = \phi_b(\theta; R)\chi_{bs}(R). \qquad (8.33)$$

This separation would be exact if the anisotropy of the potential were independent of R. The method has been called the Born–Oppenheimer Angular Radial Separation (BOARS) method. The bending wavefunctions $\phi_b(\theta; R)$ are the eigenfunctions of a rotational Hamiltonian $H_0(R)$ which depends parametrically on R:

$$H_0(R) = \frac{l^2}{2\mu_{A,XY}R^2} + \frac{j^2}{2\mu_{XY}r^2} + V(R, \theta), \qquad (8.34)$$

i.e.

$$H_0(R)\phi_b(\theta; R) = U_b(R)\phi_b(\theta; R). \qquad (8.35)$$

If this equation is solved for a series of values of R, one obtains a set of effective potentials $U_b(R)$ for the radial motion with respect to R. A zero-order wavefunction $\chi_{bs}^0(R)$ and energies E_{bs} for the radial motion are then obtained by solving the one-dimensional Schrödinger equation

$$\left(-\frac{h^2}{8\pi^2\mu} \frac{d^2}{dR^2} + U_b(R) + T_b(R) \right)\chi_{bs}^0(R) = E_{bs}\chi_{bs}^0(R), \qquad (8.36)$$

where T_b is analogous to the diagonal correction for nuclear motion in the Born–Oppenheimer separation (the term B_{ii} in equation 1.16). The dependence of the spectroscopic observables on the quantum number J can be taken into account by perturbation theory.

8.3.2. A Potential Surface for Ar–H$_2$

The infra-red spectrum of the Ar–H$_2$ molecule has been observed by McKellar and Welsh (1971) using a path length of 165 m with a pressure of 0.5 atmosphere (1 atmosphere = $1.013\,25 \times 10^5\,N\,m^{-2}$) at 86 K. They were, for example, able to observe the $S_1(0)$ transition in which j, the rotational quantum number for H$_2$, changes from 0 to 2 at the same time as $\Delta v = 1$ for the H$_2$ vibrational quantum number. Associated with each value of j for H$_2$ there is a set of levels characterized by the quantum number l for rotation of the complex as a whole. If the rotation of H$_2$ is completely free, there is no interaction between j and l but if there is a weak interaction, the level for each l value is split into several levels, each characterized by a quantum number J which can take the range of values

$$J = j+l, j+l-1, \ldots, |j-l|. \tag{8.37}$$

Information about the dependence of the potential on the parameter ξ can be obtained from transitions in which the vibrational quantum number v for H$_2$ undergoes a change. Information about the spherical part of the potential can be obtained from the $Q_1(0)$ transition for which $\Delta j = 0$ and $\Delta v = 1$. This is because the wavefunction for H$_2(j = 0)$ is spherically symmetrical and it is the spherical part of the potential that will be most important in determining the energy levels of the complex.

Potential functions of the form of equation 8.27 have been derived for the Ar–H$_2$ molecule by LeRoy and his coworkers (see LeRoy and Carley 1980). The adjustable parameters in the potential function are varied in a non-linear least squares calculation in order to get the best fit between the observed spectrum and that calculated from the potential function using the secular equation method referred to above. It is important that the potential function should be consistent with experimental data other than the spectroscopic data which only samples a limited range of R values. Although the function

$$V(R, \xi, \theta) = C_n R^{-n}(1+s_1\xi) - C_6 R^{-6}(1+s_2\xi)$$
$$+ [a_n C_n R^{-n}(1+s_3\xi) - a_6 C_6 R^{-6}(1+s_4\xi)] P_2(\cos\theta) \tag{8.38}$$

was able to account satisfactorily for the spectrum of Ar–H$_2$ and Ar–D$_2$, the values of C_6 obtained were inconsistent with the known values of the dispersion interaction. In subsequent work, the potential was expanded as in equation 8.27 with the radial functions $V_{k\lambda}(R)$ expressed in terms of a generalized Buckingham–Corner potential

$$V_{k\lambda}(R) = A_{k\lambda} \exp(-\beta_{k\lambda}R) - [C_6^{k\lambda}R^{-6} + C_8^{k\lambda}R^{-8}]D(R). \tag{8.39}$$

$D(R)$ is a damping function which gives an approximate correction for the effect of overlap on the attractive interaction. The constants $A_{k\lambda}$ and $C_8^{k\lambda}$

are functions of the parameter $\beta_{k\lambda}$, the coefficient $C_6^{k\lambda}$ and $\epsilon^{k\lambda}$ and $R_m^{k\lambda}$, which are the well depth and R_m value for each function $V_{k\lambda}$.

$$A_{k\lambda} = \frac{\exp(\beta_{k\lambda} R_m^{k\lambda})[8\epsilon^{k\lambda} - 2C_6^{k\lambda}(R_m^{k\lambda})^{-6}]}{\beta_{k\lambda} R_m^{k\lambda} - 8} \tag{8.40}$$

and

$$C_8^{k\lambda} = \frac{[\beta_{k\lambda} R_m^{k\lambda} \epsilon^{k\lambda} + (6 - \beta_{k\lambda} R_m^{k\lambda})C_6^{k\lambda}(R_m^{k\lambda})^{-6}](R_m^{k\lambda})^8}{\beta_{k\lambda} R_m^{k\lambda} - 8}. \tag{8.41}$$

Different values of $R_m^{k\lambda}$ and $\epsilon^{k\lambda}$ are allowed for each term $V_{k\lambda}(R)$. Correct long-range behaviour is assured by using theoretical values for the dispersion coefficients $C_6^{k\lambda}$ rather than determining their values in the fitting procedure. The variable parameters are $R_m^{k\lambda}$, $\epsilon^{k\lambda}$ and $\beta_{k\lambda}$. If the maximum value of λ is restricted to unity, the potential surface does not have the correct behaviour as r, the H–H distance, tends to zero. However the surface does have the correct behaviour for values of R larger than the turning point ($\approx$ 3.2 Å). Correct behaviour as $r \to 0$ can be obtained by including terms up to and including $\lambda = 3$. Differential and integral cross-sections for the scattering of H_2 by Ar can be interpreted in terms of a spherically averaged one dimensional potential between Ar and ground state H_2. The potential derived from the scattering studies is in good agreement with the vibrationally averaged spherical part of the spectroscopic potential.

8.3.3. *A Potential Surface for ArHCl*

Very accurate spectroscopic data for van der Waals molecules in the ground vibrational state can be obtained by the molecular beam electric resonance technique (Dyke and Muenter 1975). The beam source is a supersonic nozzle of the type discussed in section 7.1.1.1 and if, for example, a mixture of 3% of HCl in argon at 3 atmospheres is expanded through a nozzle, the resulting beam contains a proportion of ArHCl molecules. The beam then passes through three electric fields before reaching a mass spectrometric detector. The first and last fields (A and B) are inhomogeneous whereas the middle field (C) is homogeneous (figure 8.5). By choosing the appropriate values for the fields in the A and B regions, molecules having a particular set of J and M_J quantum numbers can be focused on the detector (cf. the discussion in section 7.1.1.1 of electrostatic fields for state selection). Microwave or radiofrequency radiation is applied in the C region and if the frequency corresponds to a transition of the molecule, the molecule will no longer be focused on to the detector by the B field. Thus absorption of radiation by the molecule is indicated by a loss of intensity at the detector. Spectroscopic data obtained by this method for a number of van der Waals molecules has

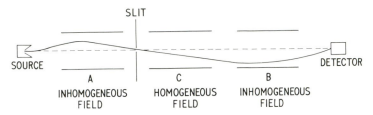

SLIT

SOURCE   DETECTOR

A   C   B

INHOMOGENEOUS   HOMOGENEOUS   INHOMOGENEOUS
FIELD   FIELD   FIELD

Figure 8.5. A schematic diagram of a molecular beam electric resonance spectrometer.

been summarized by Klemperer (1977). The rotational constant B and centrifugal distortion constant D_J can be obtained from the microwave spectrum. The rotational constant is largely determined by the static moment of inertia of the molecule and is related to the equilibrium value R_e of R. The centrifugal distortion constant D_J depends on the radial curvature of the well near the potential minimum. If radiofrequency radiation is used, transitions occur between sub-levels having different values of M_J for a given value of J. The dipole moment and quadrupole coupling constant can be derived from the radiofrequency spectrum. The dipole moment of a van der Waals molecule is largely due to the projection of the diatomic dipole moment on the the inertial a-axis of the complex. The measured dipole moment μ_a is related to the diatomic dipole μ_0 by

$$\mu_a = \mu_0 \langle P_1(\cos\theta) \rangle, \tag{8.42}$$

where θ is the angle between the inertial axis and μ_0. The measured quadrupole coupling constant $(eQq)_a$ is similarly given by the projection of the diatomic quadrupole coupling constant $(eQq)_{XY}$ on to the a-axis:

$$(eQq)_a = (eQq)_{XY} \langle P_2(\cos\theta) \rangle. \tag{8.43}$$

Because of polarization effects, data for the dipole moment are less accurate than for the quadrupole coupling constant. Thus the value of $\langle P_2(\cos\theta) \rangle$ is more accurate than that of $\langle P_1(\cos\theta) \rangle$ determined from the dipole moment.

It is also possible to measure the centrifugal distortion of the nuclear quadrupole coupling constant which can be written

$$(eQq)_J = (eQq)_{J=0} + D_Q J(J+1). \tag{8.44}$$

The average value of $P_2(\cos\theta)$ also depends on J:

$$\langle P_2(\cos\theta) \rangle_J = \langle P_2(\cos\theta) \rangle_{J=0} + D_\theta J(J+1) \tag{8.45}$$

with

$$D_Q = D_\theta (eQq)_{XY}. \tag{8.46}$$

Determination of the centrifugal distortion of the nuclear quadrupole coupling constant gives information about the coupling between angular motion and radial motion near the equilibrium geometry.

ArHCl is a prolate asymmetric top for which the asymmetry parameter b is less than 10^{-5}. The equilibrium geometry is a linear ArHCl structure with an ArCl distance of 4 Å and the HCl moiety performs large amplitude oscillations about the ArCl axis with a wavenumber of $32.2\,\mathrm{cm}^{-1}$. For ArHCl a value of $\theta = 41.5°$ is derived from the average value of $\langle \cos^2 \theta \rangle$ obtained from the quadrupole coupling constant. The amplitude of this motion is smaller in ArDCl because of the difference in rotational constants for HCl and DCl. The corresponding value of θ for ArDCl is $33.5°$. The calculation of potential surfaces for rare gas–hydrogen halide (HX) van der Waals molecules has been considered in detail by Hutson and Howard (1980, 1981, 1982). They have extended the BOARS treatment outlined above to include non-adiabatic corrections and obtain values of the spectroscopic constants B, D_J, D_θ and of $\langle P_2(\cos\theta)\rangle$ by perturbation theory. The potential for ArHCl (Hutson and Howard 1981) also fits rotational line broadening data, second virial coefficient data and the isotropic well depth determined from the integral and differential cross-sections. Since one is only concerned with the ground vibrational state, it is not necessary to consider the dependence of the potential on the interatomic distance r of the HX molecule and the anisotropic potential can be written

$$V(R,\theta) = \sum_k V_k(R)P_k(\cos\theta). \tag{8.47}$$

The Lennard-Jones and Buckingham potentials were found to be insufficiently flexible and the radial potential functions $V_k(R)$ were represented by the Maitland–Smith potential

$$V(R) = \frac{\epsilon}{n-6}\left[6\left(\frac{R_\mathrm{m}}{R}\right)^n - n\left(\frac{R_\mathrm{m}}{R}\right)^6\right], \tag{8.48}$$

where ϵ is the well depth and R_m is the value of R at the potential minimum. This function is more flexible than the Lennard-Jones or Buckingham potentials because the exponent n for the repulsive part of the potential varies with R thus

$$n = m + \gamma\left(\frac{R}{R_\mathrm{m}} - 1\right), \tag{8.49}$$

where m and γ are constants. The variable parameters in the potential are ϵ, R_m and m. If terms up to $k = 3$ are included in equation 8.47, then 12 parameters are required to determine the potential. There are not sufficient experimental data to do this so the number of parameters was reduced to 8. However, the fitting of the experimental spectroscopic data to a potential defined by equations 8.47 and 8.48 yields parameters which are highly correlated.

Hutson and Howard (1981) proposed two methods to overcome this problem. One approach expresses the potential at angles $\theta = 0°$, $60°$, $120°$ and $180°$ as functions of R using the Maitland–Smith potential. The functions $V_k(R)$ $(k = 0$–$3)$ are then obtained by interpolation using Legendre polynomials $P_0(\cos\theta)$ to $P_3(\cos\theta)$. However, this method may result in spurious angular oscillations in the potential beyond those sampled by the spectroscopic data. This can be overcome by adopting a scheme in which the parameters in the Maitland–Smith potential are angle dependent:

$$R_{\mathrm{m}}(\theta) = R_{\mathrm{m}}(0°) + \Delta R f(\theta), \tag{8.50}$$

$$m(\theta) = m(0°) + \Delta m f(\theta). \tag{8.51}$$

The function $f(\theta)$ is written

$$f(\theta) = 1 - [\tfrac{1}{2}(1+\cos\theta)]^\kappa, \tag{8.52}$$

where κ is a parameter to be optimized separately. The well depth is similarly expressed in terms of the angle θ by

$$\epsilon(\theta) = \epsilon(0°) + \Delta\epsilon(1 - [\tfrac{1}{2}\{1+\cos\theta\}]^\lambda). \tag{8.53}$$

The parameter λ is optimized independently of κ. This form of parameterization enables one to fit the potential to the angular curvature of the well near $\theta = 0°$. Equation 8.53 is independent of θ near $\theta = 180°$ and cannot reproduce a second minimum in this region. However the experimental data available at present do not give any information about this region of the potential. The calculated potential is not of the form of that in equation 8.47 but the spherical components $V_k(R)$ can be recovered by evaluating the integrals

$$V_k(R) = \int_{-1}^{1} (k + \tfrac{1}{2}) V(R, \theta) P_k(\cos\theta)\, \mathrm{d}\cos\theta. \tag{8.54}$$

It may be necessary to include up to 10 Legendre polynomials in order to obtain an accurate representation of $V(R, \theta)$ in the repulsive region. The potential surface obtained by this method for ArHCl is shown in figure 8.6 in which the potential minimum for $\theta = 0°$ is clearly visible.

Spectroscopic constants calculated using this potential surface are in very good agreement with the molecular beam electric resonance data. Scattering calculations show that this potential can predict accurately the position of the rainbow maximum and the magnitude of the high-angle inelastic scattering. However, an earlier potential surface, which included a secondary minimum at $\theta = 180°$, was more successful in predicting the amplitude of the rainbow oscillations. This suggests that there may be a secondary minimum at $\theta = 180°$. In the case of NeHCl, the barrier to internal rotation for HCl is comparable with the rotational constant of HCl. Thus the motion of HCl in NeHCl is better described as a hindered rotation and the spectra

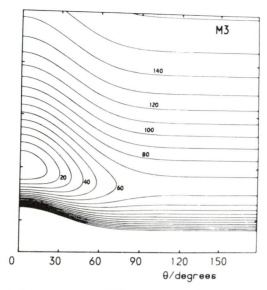

Figure 8.6. A potential surface for ArHCl. (Reproduced from Hutson and Howard 1981.)

are sensitive to the whole range of θ. The method outlined above has been modified (Hutson and Howard 1982) so that the resulting potential surface is capable of describing a minimum at $\theta = 180°$. A potential surface has been calculated for NeHCl (figure 8.7) and it has a secondary minimum at $\theta = 180°$. Use of this more flexible form of a potential function for ArHCl yields a potential surface which is capable of fitting the spectroscopic data to the same accuracy as before as well as predicting a secondary minimum at $\theta = 180°$.

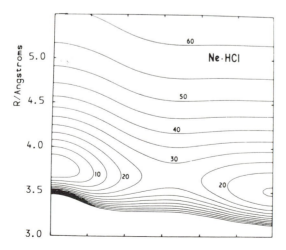

Figure 8.7. A potential surface for NeHCl. (Reproduced from Hutson and Howard 1982.)

Suggestions for Further Reading

B. L. Blaney and G. E. Ewing (1976) *Ann. Rev. Phys. Chem.*, **27**, 553.

G. E. Ewing (1975) *Acc. Chem. Research*, **8**, 185.

B. J. Howard (1975) in *International Review of Science, Physical Chemistry, Series 2,* Vol. 2, ed. A. D. Buckingham, p. 93, Butterworths, London.

References

P. R. Certain and L. W. Bruch (1972) in *M. T. P. International Review of Science, Physical Chemistry, Series 1,* Volume 1, ed. W. Byers Brown, p. 113, Butterworths, London.

E. A. Colbourn and A. E. Douglas (1976) *J. Chem. Phys.*, **65**, 1741.

D. E. Freeman, K. Yoshino and Y. Tanaka (1974) *J. Chem. Phys.*, **61**, 4880.

B. J. Howard (1975) in *International Review of Science, Physical Chemistry, Series 2,* Volume 2, ed. A. D. Buckingham, p. 93, Butterworths, London.

J. M. Hutson and B. J. Howard (1980) *Mol. Phys.*, **41**, 1123.

J. M. Hutson and B. J. Howard (1981) *Mol. Phys.*, **43**, 493.

J. M. Hutson and B. J. Howard (1982) *Mol. Phys.*, **45**, 769.

W. Klemperer (1977) *Farad. Disc. Chem. Soc.*, **62**, 179.

W. Kutzelnigg (1977) *Farad. Disc. Chem. Soc.*, **62**, 185.

R. J. LeRoy (1973) in *Molecular Spectroscopy*, Vol. 1, ed. R. F. Barrow, D. A. Long and D. J. Millen, p. 113, The Chemical Society, London.

R. J. LeRoy and J. S. Carley (1980) in *Potential Energy Surfaces,* ed. K. P. Lawley, p. 353, Wiley, Chichester.

R. J. LeRoy and W.-H. Lam (1980) *Chem. Phys. Lett.*, **71**, 544.

D. H. Levy (1981) in *Photoselective Chemistry*, Part 1, ed. J. Jortner, R. D. Levine and S. A. Rice, p. 323, Wiley, New York.

A. R. W. McKellar and H. L. Welsh (1971) *J. Chem. Phys.*, **55**, 595.

G. C. Maitland, M. Rigby, E. B. Smith and W. A. Wakeham (1981) *Intermolecular Forces,* Clarendon Press, Oxford.

T. R. Dyke and J. S. Muenter (1975) in *International Review of Science, Physical Chemistry, Series 2,* Volume 2, ed. A. D. Buckingham, p. 27, Butterworths, London.

J. A. Pople (1982) *Farad. Disc. Chem. Soc.*, **73**, 7.

J. G. Stamper (1975) in *Theoretical Chemistry*, Vol. 2, ed. R. N. Dixon and C. Thomson, The Chemical Society, London.

INDEX

229